Newham London

ADULTS

up

24 hour automated telephone renewal line
0115 929 3388
Or online at www.newham.gov.uk

This book must be returned (or its issue renewed)
on or before the date stamped above

THE
HOVERCRAFT
A HISTORY

ASHLEY HOLLEBONE

First published 2012

The History Press
The Mill, Brimscombe Port
Stroud, Gloucestershire, GL5 2QG
www.thehistorypress.co.uk

British Library Cataloguing in Publication Data.
A catalogue record for this book is available from the British Library.

ISBN 978 0 7524 6479 4

Typesetting and origination by The History Press
Printed in Great Britain

CONTENTS

Foreword by Warwick Jacobs   7

Introduction   9

1   Where it all Began   11

2   The Workings of a Hovercraft   27

3   SRN-1: The First Hovercraft   31

4   The Next Chapter…   37

5   The British Hovercraft Corporation
    and the SRN-5/6: Hovering the World   69

6   BHC SRN-4 Mountbatten Class: The 'Super 4'   79

7   Hoverspeed   105

8   Military Hovercraft   110

9   Wing in Ground Effect: The Age of the Ekranoplan   133

10 Light Hovercraft   140

11 Hovertravel: The World's Oldest Commuter Hovercraft Service   153

12 Griffon Hoverwork   164

13 The Future   177

Appendix 1   Patents of Sir Christopher Cockerell   180

Appendix 2   Griffon Hoverwork Sales List, 1983–2012   184

Index   188

FOREWORD

It is not every day you get asked to write a foreword for a book, and indeed I wasn't actually asked; I was tasked with asking either the hovercraft inventor's daughter, Frances Cockerell, or Lord Mountbatten's grandson, Lord Romsey, both being Patrons of the Hovercraft Museum Trust of which I am a founding trustee. Needless to say Frances Cockerell was invaluable and preferred to be involved in helping to proofread and get the Cockerell facts right as indeed Sir Christopher would have wanted. Though the book doesn't emphasise this, he received very little remuneration for his fantastic invention; a knighthood and the equivalent of a deposit on a nice little bungalow. It certainly wasn't a living and he had to go on inventing, with wave power and future energy alternatives being next. Had he been an American he would have been a millionaire, though Cockerell never complained.

If it wasn't for Dr Cockerell and Lord Louis Mountbatten's promotion of his work then maybe 'Hovercraft' would have been the American 'Ground Effect Machine', or the French 'Aeroglisseur', or the Austrian 'Leufkissenboot' – maybe even the 'Vazsnashino' from the Eastern Bloc. This very British invention nearly didn't happen except for the perseverance of these two great Englishmen.

So for me it is with great pleasure that I get to write the foreword for Ashley's hovercraft history. Like Ashley and many a schoolboy I fell in love with hovercraft at first sight. With great engineering, the hovercraft is the ultimate transformer machine as it metamorphs as it inflates, lifts and drops, changing from sea to land and back. Always a crowd puller with its speed, spray and noise – rudders and props moving and skirts bellowing – the hovercraft is a quirky, futuristic sci-fi machine.

When I first met Ashley he was a student who turned up at an early hovershow asking if he could bring in his rare American Hoverstar craft towed by an Austin Seven car, driven down from London. Needless to say we said yes and enjoyed his display only to find afterwards that he'd never even driven it before!

Ashley's enthusiasm comes through in his book and I hope that this will introduce a new generation to the hovering craft, as well as educate us older 'schoolboys'! As Cockerell always said and believed, it is the youth of today we have to enthuse as they are the future and we have to educate people on both the Arts and Sciences or else we are half educated.

Cockerell was my idol, and I had the pleasure of meeting him aged thirteen. Years later I was to paint his portrait whilst he quizzed me on the History of Art and his latest letters to *The Times*. He was a well-rounded gentleman and knew his stuff! He would have liked Ashley (writer, actor, journalist and props expert on the Bond films) and I know his book will go a long way to educating the next generation of engineers and enthusiasts on this wonderful machine. As Cockerell said himself, 'Oh well you have to understand it is only half developed you know ... you can't un-invent something. There will always be Hovercraft!'

With this book, Ashley brings us right up to date and ready for the future!

Warwick Jacobs
Trustee, Hovercraft Museum Trust, Lee-on-the-Solent

INTRODUCTION

In 1959 a new invention came into being, and it wasn't long before it changed the world, inspiring many designers, visionaries, businessmen and members of the public. This was the hovercraft, and this is my account which chronicles the history of this wonderful invention that is statistically one of the world's safest modes of transport. The hovercraft is having something of a renaissance, growing from strength to strength as it continues to meet the world's demands for an amphibious and truly versatile vehicle. Nothing else can match it!

I would like to express my gratitude to Hovercraft Museum trustee Warwick Jacobs, not only for his hard and loyal work devoted to the preservation of countless historic hovercraft, but also for the archiving of tens of thousands of images and cuttings that span the entire evolution of the hovercraft.

A home-made 1960s hovercar which used the running gear and wheels from a car to provide an adequate means of control for the road. This strange vehicle was also road legal, utilising the registration from the car which it was mechanically based upon.

The Hovercraft Museum near Gosport, Hampshire, on the south coast of England houses a truly unique collection of the most important hovercraft and archives to be found anywhere in the world. (Author's collection)

This book is my opinion on the hovercraft and I really wanted to finally make good use of many images that have never been seen before, all of which help to document and appreciate the history of the hovercraft, and of the men and women that pushed it to the extremes and gave the world a life-saving vehicle. I hope that you enjoy gaining an insight into this unique mode of transport.

Unless otherwise stated, all images come from the collection of the Hovercraft Museum.

ONE

WHERE IT ALL BEGAN

Six thousand years ago, during the fourth millennium B.C., modern man's hairy ancestors were formulating the idea of a wheel hewn from stone, which they later made from wood, having discovered this material was easier to work with and could be more durable. They had proved that moving objects could be made easier by a simple method, one that would enable them to advance their lives and civilisations. Six thousand years later, and only a short time after Elvis had a hit with *Hound Dog*, another strange but forward-thinking concept would emerge in the twentieth century. By now the human population of the world was familiar with the aeroplane in both military and civilian applications. Cars were continuing to get faster and faster, ships were getting bigger and bigger, while trains were slowly turning from imperial steam to diesel and electric traction.

Earth is a wonderful place but few of us ever really take note of the natural beauty that is all around us, or at least not far from wherever we may be. Our planet is covered by a thick blanket of air which is over 200 miles in depth, commonly referred to as the atmosphere, although there are different areas to be found within this zone. These include the stratosphere, jet stream and ionosphere, all of which vary in air pressure; the higher you rise, the lower the pressure of air and thus the lower the amount of energy required to achieve high speeds as the earth spins on its axis below. High altitude jets, such as the now sadly redundant Concorde, ferried their passengers on the very edge of space (58,000ft), while satellites glide even further above us as we sleep.

We have three main elements that make up our environment, these being land, sea and air. Man has achieved transportation through all of these media but only after many years of development. Since the innovation of the wheel, the boat is the oldest form and records can prove that thousands of years have passed since the first examples were constructed. The aeroplane is considered a definite twentieth-century breakthrough, though medieval genius Leonardo da Vinci proved that powered flight was a definite possibility. So on the face of it, transport history would seem to suggest that these three elements have been dealt with, and that man has firmly mechanised earth. Of course, as with all things, it is only a matter of time before someone will try to improve an existing creation or in some cases form a totally new one.

If we examine the three modes of transport we will find flaws in all of them. Ships are the most commercially used mode of transport but are slow; in most cases the maximum speed of a cargo ship is not more than that of the top speed of a 100-year-old motor car! Where an increase in power is obtained it does not translate into forward momentum as that extra power is needed to punch through the water. Boats in their state of movement require quite a lot of energy and high-speed vessels have also presented efficiency challenges. A ship is a much larger version of a boat which may be used for private and recreational use, although in both cases these vessels have their limits, requiring docks or dredged channels to berth.

Wheels are the most practical for everyday day use but they too have limitations. Trucks can carry heavy loads but nowhere near the quantity a ship can carry, while the size of load is constricted by the roads on which it will travel. Further to this, they are heavily affected by traffic congestion, a problem not suffered to any great extent in the air or at sea. The wheel is also greatly affected by severe weather conditions; when placed under stress a wheel can slip in icy conditions and lose all traction. Snow chains and desert ballon tyres provide a practical solution in some cases but the drawbacks to the wheel in extreme conditions are far greater than can be rectified with aftermarket creations for the masses.

Flying is by far the fastest way to get around but also attracts huge costs. Aircraft only stay airborne at speed and require large amounts of energy to get off the ground, as well as to stop! Generally aircraft loads are quite small by comparison to other forms of transport at smaller costs.

To be quite blunt, the faster you travel the more it costs. As the ever-increasing world population demands its resources and consumer items at an ever-increasing pace in the manner of 'I want that and I want it now!' then so does the need follow for speed in delivery. The aircraft being too costly, shipping being too slow and the wheel being limited to land it was clear that a gap existed.

By the middle of the twentieth century an unorthodox-looking invention entered public knowledge; however despite its appearance it did not emerge from a secret government test site in the Nevada Desert or from a Cold War British air base but a small, rustic boatyard in Norfolk, the same place where turnips and mustard originate, and later plastic sports cars! However, while the creation's home may have been simplistic, its qualities for the future were anything but. This was to become quite possibly one of most futuristic-looking vehicles of the post-war industrial era, the hovercraft.

As with all forms of movement, friction is the inevitable barrier that has to be overcome. The hovercraft is based around the idea of eliminating friction, and while the hovercraft itself did not appear until the middle of the Cold War era, ideas of transport focusing on overcoming friction originated centuries earlier.

Whilst da Vinci's concept of a human-powered flying machine never made it past his famous sketch, it paved the way for future aviators. Da Vinci's flying machine has in later years been built by countless historians who have proved that this medieval genius was in fact the inventor of the helicopter, although it was aviation firm Sikorsky that would make it a reality some half a millennium later. The theory of hovering can therefore be traced back to the artistic, rapidly advancing Renaissance era.

The next chapter in the air cushion history comes from the early eighteenth century when, in 1716, Swedish philosopher and designer Emmanuel Swedenborg devised an air-cushioned vessel which resembled an upturned dinghy with a cockpit in the centre. Apertures on either side of this allowed the operator to raise or lower a pair of oar-like air scoops, which on the downward strokes would force compressed air beneath the hull, thus raising it above the surface. The project was short-lived and it was never built, for Swedenborg soon realised that to operate such a machine required a source of energy far greater than that which could be supplied by a single human occupant.

Hovermarine sidewall vessel – note the lack of a bow wave.

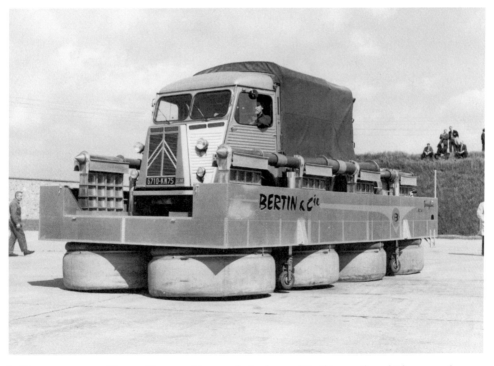

In France, not even a Citroen H van could escape being hovered! In this case the vehicle was used as a load to demonstrate the advantages of an air cushion pallet.

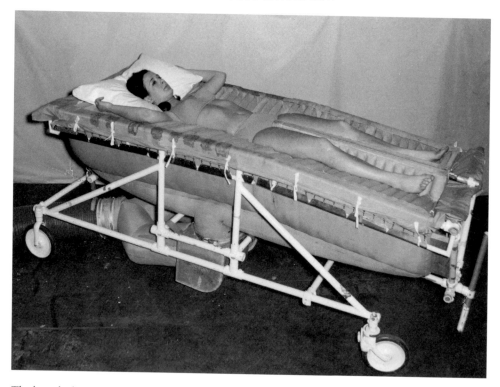

The hoverbed. A British invention that helped and continues to help the treatment of major burn victims.

This strange boat is actually one of the first full-scale working examples of an air cushion vehicle. The 1916 'Gleiftboat' was a high-speed German torpedo launch that used an elaborate air blower system to reduce friction under the hull while its shape meant the bow would rise out of the water at speed. A design way ahead of its time.

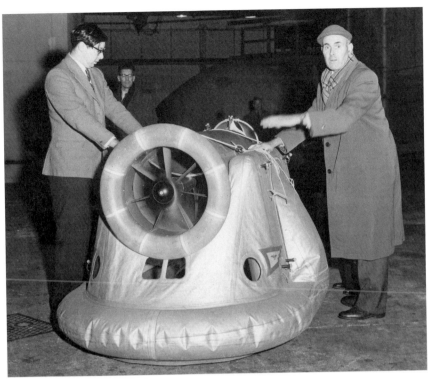

A light one-person hovercraft made at the Royal Aeronautical Establishment at Cardington, Bedfordshire.

The hovercraft at Cardington. From this shot it is clear to see why the base was used to experiment in hovercraft activities. The building is one of the old airship hangars which once housed the famous *R101*, however in the background lie barrage balloons, perhaps surplus from the Second World War?

Another RAE Cardington inflatable craft.

'The Skimmer', 24 April 1961, under the control of its designer, former test pilot for Supermarine (manufacturers of the Spitfire) Don Robertson. Whilst trying not to look like he is sat on the toilet, Don skilfully hovers the craft under the watching eye of the press.

Winfields hovering at Jersey Hoverdrome in the 1970s.

Through the next century, things progressed with a range of practical experiments with air cushion transport. This time the location was England in the 1870s, when engineer John Isaac Thornycroft experimented by forcing air underneath the hull of a small boat. Thornycroft was no stranger to the water as he was already involved in constructing boats for the Admiralty from his yard on the Thames. Thornycroft's idea was to reduce friction of the hull as it passed through water, enabling a reduction in drag so an increase in momentum. He constructed a form of crude bellows system which was pumped from inside, drawing air from the top and then vented to the underside of the hull. The idea worked, as air bubbles formed and displaced water. However, the power source required to make this method a viable proposition just wasn't around at that time. It wouldn't be until a century later that the frictionless vehicle idea would 'surface' once again. Thornycroft died in 1928 but his legacy would remain to the end of the twentieth century as his firm went on to build the famous RAF motorboats for air rescue duties during the Second World War. The Thornycroft name was also used for commercial motor vehicle production, most famously with fire engines and trucks. A great number of Thornycroft's original working models have thankfully been preserved and can be seen on display at the Hovercraft Museum in Hampshire.

Now we reach the final chapter in the experimental phase of the concept of air cushion theory when in 1955 another highly skilled but very much independent man of modest means transformed centuries of experiments into a practical product. Dr Christopher Cockerell (later knighted) is known as the inventor of the machine we call the hovercraft, a word he also coined, but few are aware of the man himself. To appreciate invention you have to appreciate and understand the inventor and when you delve into Cockerell's life history it's not hard to see why it would be such a character that would create this breakthrough.

Born on 4 June 1910, Cockerell was part of a very well-placed family in society. His father, Sir Sydney Carlyle Cockerell, was a private secretary to Sir William Morris and from 1908–37 was Director of the Fitzwilliam Museum in Cambridge.

The Cockerells were a talented family. Sir Sydney's parents were Sydney John Cockerell, a London coal merchant, and Alice Bennett, the daughter of a City watchmaker, while his elder brother, Theodore, was a biologist. His younger brother, Douglas, an eminent bookbinder, had

Inventor of the hovercraft as we know it, the English gentleman Sir Christopher Cockerell.

Christopher Cockerell is honoured by HRH the Queen Mother at Buckingham Palace in 1969.

a son, Sydney Maurice, who was two years Cockerell's senior and a celebrated and innovative designer of marbled papers.

Cockerell studied engineering at Peterhouse, the oldest and smallest college of Cambridge. It seems a strange turn of events that such a historic old city would lay claim to the world's most futuristic contraption of the decade, if not the century! After his studies at Cambridge, Cockerell gained employment at the Radio Research Company until 1935 when he moved to Marconi. He stayed with the wireless telegram firm until 1950.

Cockerell's father once described his son as 'no better than a garage hand'; despite this he had a vast capacity for invention. Throughout his life Cockerell filed numerous patents for a wide variety of designs and inventions. He created thirty-six in his short time with Marconi alone, and despite his reservations, his father funded his son's files for many of his patents. Over history it is often the case that great people of our time can be interlinked with other such personalities. Cockerell's father was no exception as he was associated with the likes of T.E. Lawrence (Lawrence of Arabia) and George Bernard Shaw among others, who were often house guests.

During the war years Cockerell worked with a highly skilled elite team at Marconi developing radar, a development which Winston Churchill believed had a significant impact on the outcome of the Second World War. Cockerell left Marconi in 1950 and with the financial aid left by his dear wife's father, he and his wife Margaret were able to purchase a small boatyard in Suffolk. It wasn't long before they had established a new venture at the Old Wherry Dyke in Somerleyton which had in the previous century been used as a quay for a Victorian brickworks, although it was at that time home to broad cruisers which were growing in popularity. A new company was registered, Ripplecraft, through which Cockerell would run his marine business of hiring out cruising boats as the demand for holidays on the Norfolk Broads increased. Each year Cockerell would plan to lay a new keel over the winter period so that a new boat would be ready for hire in the following summer. There were alterations made to the conventional Broadland cruiser as Cockerell insisted on placing the steering wheel at the front in a car-like fashion so that hire customers could enjoy the panoramic views of the countryside when the large sliding roofs were open. Ripplecraft was an active and vibrant little boatyard and the small fleet of Broadland cruisers allowed Cockerell to provide a living for his family, as well as to fund his development work on new technology.

He didn't run this venture alone, however; behind the scenes were other highly skilled and very much underrated men that had great experience in design, construction and operation of boats. One such man was Douglas Rushmer, who remained managing director of Ripplecraft until 1979. Doug worked very closely with Cockerell on the testing of various air cushion ideas, including a side wall concept which utilised a modified clinker-built rowing boat. Thin side members were added which penetrated the water line. At the keel a transverse slot was cut from where air would enter under the bottom of the craft. An industrial vacuum cleaner was used to provide the required air pressure. The results proved that friction had decreased and this led Cockerell to carry on and eventually purchase a small motor launch named *Spray* that he could use to develop the idea to another level under powered motion.

Spray was modified with a large centrifugal air fan that was powered from its propeller shaft. Air was vented over the bows of the launch and to determine the pressure pattern a series of simple water height gauges were installed through the bottom and could easily be monitored.

But despite all of this Cockerell still had his day-to-day business to run and playing with toy boats like boys on the riverbank on the way home from school does not pay the bills. All of this kept Cockerell busy, but his brain was buzzing for further creation and during the winter of 1953 he started to give his air cushion idea some serious thought.

His experiments led him to construct large-scale models which would test his idea about making a boat ride upon a cushion of air. Cockerell spent much time in his quest to pursue a frictionless craft. One example of his experiments involved a small dinghy which had a special pump to blow high-pressure air underneath and around the edge of the hull. This high pressure air was then retained by the aid of a rubber curtain which then created lift. This working test bed was to become the first step in the development of the hovercraft as we know it. It was one Saturday evening in June 1954 that Cockerell stumbled upon a method that would actually turn all of his hard work and forward thinking on air cushion ideas into a viable, fully working example. He had created the very item that makes a hovercraft hover – the momentum curtain. For some time he had been playing with the notion of devising some sort of contained pressured segment on the underside of the hull. Cockerell thought of having a plenum chamber; a 'storage' chamber where the entering air would be contained for a little longer, and thus pressurised, so as to reduce the power needed for given lifting capacity.

Making good use of what he had around him, Cockerell used two tin cans of slightly different circumferences: a Lyons coffee tin and a Kitekat cat food tin. The smaller tin was recessed within the larger one and fixed at the top, leaving a gap around the sides. A hole was then made in the top of the large tin and from there an industrial air blower hose was connected. The result created a desired plenum chamber as the high-pressure air entering the tin was forced to dissipate around the sides of the smaller tin, exiting at a greater pressure from the underside of these two cans. This method, seemingly so simple, created a highly effective air curtain and test model that would go on to prove the fundamental basis of the future of the hovercraft. This was called the 'momentum curtain' and Cockerell filed yet another patent, one of his most significant to date, although it made him very little money.

Cockerell used this simple coffee tin experiment to demonstrate the principle and goings-on of the plenum chamber. The cans were made in such a way that it was possible for the smaller inside can to be removed so that the exiting thrust air could be directed onto a pair of scales with weights on. The test showed that the scales would not move much when the tin was placed over the scale; however, when the smaller tin was reinserted, a plenum chamber was created and the tin rig offered up to the scale once more. This time the weights lifted, showing a clear advantage in air having the momentum curtain as for the same amount of power a greater exiting force was reached. The test rig proved highly successful. Now all Cockerell needed to do was build a working model that would show this in a more practical and suitable manner.

It is quite easy to reproduce this simple test yourself using basic household items just like Cockerell did himself. You can even use plastic drinks bottles.

A full working model of the concept was built, which resembled a cross between a flying helmet from a Dan Dare adventure comic and a spaceship. Yet this balsa wood contraption was powered by a small model glow-plug aero engine. Even more primitive, compared to its futuristic appearance, was the fact that it was tested on bowling greens on a tethered rope. The design was of Cockerell's mind but the construction wasn't by his hand. He simply did not have the time to make models whilst managing the growing boatyard, so he called on the skills of his colleague Desmond Truman to build the working model of this revolutionary design. This model was Cockerell's first working example which he would use to prove his air cushion theory to potential investors. Like the models of John Thornycroft of the Victorian era, Cockerell's original is preserved at the Hovercraft Museum. It was at this time that Christopher Cockerell gave his new design the name 'hovercraft'.

Another company was formed alongside Ripplecraft, Hovercraft Ltd. The small ramshackle old shed which had been used in the past to store unused tools and boat bits became the workshop for the most iconic vehicle of the decade if not the century, similar to the Apollo rockets. But of course, this was not a multi-billion dollar NASA plant, this was the Norfolk countryside.

The impressive stately home of Somerleyton Hall lay within a stone's throw from the Ripplecraft yard and in June 1956 Cockerell met with Lord Somerleyton, who granted him full permission to use the estate to further his work on the hovercraft model tests within the seclusion of the Pergola Lawn. This was an important place for Cockerell, as he could undertake tests without having to concern himself of onlookers that could place an issue. The connection with Somerleyton deepens further as it was Lord Somerleyton who initially contacted Lord Louis Mountbatten in August 1957, who was at the time First Sea Lord of the Admiralty and had a well-known interest in technology and innovation. In his forthright way he later said, 'I sent the papers on to the Admiralty Director of Research Programes and Planning with my personal instructions that this was to be investigated at once.' Mountbatten turned out to be as vital a part of the hovercraft story as the momentum curtain itself, but more on that later.

The first practical hovercraft model built by Cockerell, seen here on Lord Somerleyton's lawn in Norfolk.

Cockerell's model of the craft of the future.

Cockerell preparing the model for a tethered flight on Lord Somerleyton's lawn.

Eventually in 1955 Cockerell managed to gain the support of the Ministry of Supply but the MoD was another task in itself! He tried to persuade officials that his new hovercraft was the vehicle of the future and one that could meet certain requirements in various roles. Sadly he was met with firmly shaking heads from the armed forces. The Admiralty said that it was an aircraft and not a boat, the RAF said it was a boat and not a plane! The Army held neither of these opinions but were not interested. It is quite ironic that most of the world's military hovercraft are today operated by the Marines.

One other rather unsettling thing became clear from this early period of the development of Cockerell's craft. More or less as soon as he had patented the design of the craft and the momentum curtain, it was placed on a secret list. In an almost James Bond fashion you could picture a sunken room deep within the Whitehall citadel where numerous weird and wonderful gadgets are filed, stored and played with by mature men in Savile Row suits and bowler hats. This secret list prevented anyone other than those with direct permission from government authorities, or in some cases deeper, from doing anything with the patent or project. These were dark times, the Cold War was getting very chilly, the USSR was watching Great Britain and the United States while Great Britain and the United States were watching the USSR. There was never any doubt that a very short fuse existed that could be lit at any time, and everything relied on technology being developed by boffins at defence research establishments. During this time the UK had firmly launched its greatest deterrent, the V bomber force, which consisted of three types of high-altitude, long-range jet bombers capable of dropping the *Blue Steel* nuclear missile.

The hovercraft arrived at the peak of a major RAF overhaul; its piston-engined aircraft which dated back to the Battle of Britain were now being replaced with supersonic jets that could break the sound barrier in a vertical climb, seconds from take-off. It's not hard to see why the powers that be were determined to tread carefully with this new invention. When taken to Whitehall the model was demonstrated amongst a gathered crowd of senior top brass. It must have been a strange sight to witness this free flight little model, noisily hovering in circles over the grandest flooring it was ever likely to, a world away from a leaky damp shed in a Suffolk boatyard! The project was quickly whisked from under Cockerell and placed on the secret list, although as already mentioned not because they had a plan for it, but just to stop someone else having a plan of their own for it. Could it have assisted their nuclear deterrent? Or aided their ground forces? Only time would tell. What was clear was that Cockerell was getting increasing agitated by the suppression of the technology. Eventually in 1958, after declassification, Cockerell got a step further. A member of the Ministry of Supply was rather concerned, as was Cockerell, about reports of hovering progress being made abroad and together they saw the need to act fast in Britain to keep her foot in front as she had done so far. Mr R.A. Shaw of the Ministry of Supply gave permission to the National Research Development Council (NRDC) to fund the design, construction and development of the world's first full-size working hovercraft, the SRN-1.

The design code of SRN-1 originates from the name of the manufacturer given the contract: aircraft firm Saunders Roe – Nautical division, a company based in Cowes on the Isle of Wight, just off the south coast of England. The Isle of Wight evolves from here onwards as the birthplace of the hovercraft to being the very epicentre of its activity. Saunders Roe, meanwhile, was one of Britain's oldest aviation companies, with a long connection with vehicles that combined air and sea, having built some of the most important seaplanes throughout both world wars.

The Saunders family came from Streatley, situated on the Thames, where they designed and built boats for use on the river. They continually refined their designs and methods of construction in order to make the boats lighter and more durable. The weight and hull design became more important when engines were installed, as the early designs were not very efficient. In addition, they also made great attempts to reduce the wash generated by powered craft.

Sam Saunders established his business on the Isle of Wight in the early 1900s and soon began manufacturing boats for open sea applications in addition to his existing river boats business. The sleekness of his designs and their light weight soon brought many new customers. This was particularly important when the early aircraft were being built. After building fuselages for a number of aircraft, it was not long before Sam Saunders commenced building his own aircraft. His company was responsible for many very interesting aircraft designs.

As the business expanded, changes in the ownership and management structure of the company occurred, typical of any pioneering business, and naturally the name of the business also changed. In 1929, when Sir Alliott Verdon Roe bought into the company, Saunders Roe was born and set on its way to becoming one of the more enduring names in the UK aircraft industry. Although they became synonymous with flying boats in particular, it should not be forgotten that the company also built several land-based aeroplanes.

Sir Alliott Verdon Roe was born on 26 April 1877 in Patricroft, Lancashire. Roe was not keen to follow in the family nature of caring for others, his father being a doctor and his mother involved with children's nurseries. Roe had interests in engineering, in particular flying. He left school at the age of fourteen and amazingly went off to work for a civil engineering company in British Columbia, Canada. But it wasn't to last long and after no time Roe was back in employment, serving an apprenticeship in Portsmouth dockyard. He then went on to study, appropriately enough, marine engineering at Kings College, London. Roe also somehow managed to find the time to join the company of the SS *Inchanga* as fifth engineer. It was during this time that Roe first turned his mind to the possibility of actually building a flying machine. He made various models and even travelled back over the pond to work in the USA on a new gyrocopter, although that wasn't a great success and Roe soon came back to Blighty to pursue his interest in aviation further.

In 1906 he patented the first aircraft control column to comprise a single level, where previously two had been used. This is a device still in use to this day on light aircraft. On 8 June 1908 Roe flew for the first time and in his own aircraft from his new base at Brooklands. Unfortunately Roe did not register his achievements with The Royal Aero Club, who were even based at Brooklands! Sadly for Roe the glory of achieving the first powered aircraft that could sustain a controllable flight was to go to Lord Brabazon. Before this, aviation was made up of complex, but fragile, glued-together contraptions. These early aviators literally put their lives in the hands of God, in the hope of taking to the skies like birds. The period between the death of Queen Victoria and the outbreak of the First World War would without doubt be the dawn of what was to come; the very start of man venturing into the clouds in a manner that we understand and practise today.

The aeroplane had arrived and Roe pressed on regardless. It was a challenging time as he had to learn so much. Designing, building and flying all at once! It was dangerous because things kept breaking. Controls were not correctly understood and engines were unreliable.

After being evicted from first Brooklands and then Hackney Marshes, Roe set up his flying operations at Wembley on the outskirts of London. Roe's brother, Humphrey, who was later to marry Marie Stopes, came into the business and on New Year's Day 1910 A.V. Roe became the first company ever to be registered as an aeroplane manufacturer. The learning curve was so fast that hardly ever were two aircraft built that were exactly the same. Improvements came along at a breathtaking pace.

Manufacturing moved to Manchester and with Brooklands under new management an Avro flying school was set up there, later moving to Shoreham-by-Sea, West Sussex. Other money-making ventures were the founding of an aircraft spares warehouse and the invention and marketing of a turnbuckle for tightening the bracing wires used on aircraft in those days.

After the First World War military orders dried up to a trickle and even with new designs orders were small. The civil market was hotly contested and Avro's most successful aircraft was the Avro Avian. Even before this a considerable financial investment had been made in the company by the Groves family, of Groves and Whitehall Ltd, the Manchester brewers. In 1920 Crossley Motors bought three-fifths of the shares in the company, then in 1928 control of the company passed to the Armstrong Siddeley Development Group. As a result both of the Avro brothers, Alliott and Humphrey, left to join S.E. Saunders Ltd of Cowes, Isle of Wight. Saunders, soon to become known as Saunders Roe, were, as we have seen, exponents of the flying boat. Once known as Saunders Roe, they produced a series of well-known aircraft, culminating in the 'Saro Princess'. They also produced the very successful 'Skeeter' helicopter and the experimental rocket fighter, the SR-53. Around the same time, Alliott Verdon Roe was knighted in the New Year Honours list of 1929.

The company was also one of the pioneers in helicopter design and it was at this important point in the company's development that the British Government rationalised the UK's aircraft industry in 1957. This reorganisation basically grouped all the fixed wing manufacturers together under the name of British Aircraft Corporation and all of the helicopter builders under the Westland Aircraft name.

Saunders Roe became the Saunders Roe division of Westland Aircraft, as it was also designing helicopters in addition to fixed-wing aircraft. Sam Saunders was granted a patent for his method of hull design which was called 'consuta' where layers of relatively thin planking were glued and stitched together with copper wire making a lightweight hull which was very strong. This was an early form of composite material.

Saunders Roe would work with Christopher Cockerell's guidance on this new hovercraft project avidly over the coming months. Their first prototype model looked quite different from Cockerell's flying man helmet as it had a circular body with a large inverted duct in the middle. A cabin was placed at the front from where the craft could be operated by its minimal crew. Propulsion was provided by ducts which vented air from the large fan used to provide the lift. The model went through much testing including over land-based obstacles and proved sufficient enough to grant the next stage of development, which was to actually build the real thing.

Ahead of schedule, the 7-ton craft flew on 31 May 1959, only eight months after the commencement of design work. The 20ft craft was quickly nicknamed the 'flying saucer' and it's really not hard to see why. But it was not until 11 June that she made her first public appearance in front of the world's press. Such was the interest in this new form of transport that the press refused to leave until she was demonstrated in the water. Within weeks, on 25 July, she made a crossing of the English Channel, from Calais to Dover, with Cockerell aboard as human ballast, on the fiftieth anniversary of Blériot's first aeroplane crossing of

The Mitsubishi corporation of Japan created a hovercraft which looked very much like the British SRN-1.

the Channel. Cockerell's dream had become a reality. Since then, hovercraft have carried over 100 million passengers and 13 million vehicles and have been in service for over fifty years.

It must have been quite a shock to the crowd that gathered at Dover to watch the hovering saucer skim its way onto the beach, in a mist of sea spray and with a deafening noise. This old-fashioned seaside town, which can trace its roots back 1,000 years, was to host the landing of the SRN-1 hovercraft. Dover would later resume its hovercraft connection when a cross-Channel service opened in 1969.

TWO

THE WORKINGS OF A HOVERCRAFT

To understand further about the history of the hovercraft, it is important to look at the goings-on under her skirt. A hovercraft is probably one of the simplest forms of transport, next to the wheel which, though simple, seems to become more and more refined itself as years advance.

A hovercraft flies; it is not in contact with the ground in its operation, therefore technically its airborne flight characteristics make it an aircraft. Hovercraft, like a conventional airframe, have rudders for steering and elevators for trim and adjustment of height. In common with more classical 'aircraft', hovercraft can also have radar, artificial horizon gauges, variable pitch propellers, and bow thrusters.

Reducing the friction of the ground is the easiest way to move an object with little effort. However, hovering in itself raises issues of stability which have been addressed by effective aerodynamic control surfaces as with aeroplanes. If we look at the underside of a hovercraft it appears to be flat but look closer to the outside edge and you will notice a series of holes running around the entire craft. This cavity you are looking at is called a plenum chamber which derives from the Latin word *plenum* meaning 'full', as in this case the chamber becomes full of pressurised air.

The three most characteristic parts to a hovercraft are the plenum chamber, momentum curtain and engine. While you may think it might be the skirt, this is only an addition to increase performance as we saw with SRN-1 and other later variants. A hovercraft will still hover, regardless of whether it has a flexible skirt or not. The momentum curtain, as patented by Christopher Cockerell himself, displaces the air evenly, creating a much-needed stable platform of air pressure under the craft.

At low speeds a lot of energy is required to make sufficient progress when passing an object through water. An example of this you will be familiar with is when you are in the sea or swimming pool; as you walk through the water, you quickly notice how hard it is for your muscles to fight the water resistance. Water aerobics is a common form of exercise, making use of the volume of water to stay fit. The same principle of increased effort applies for boats.

When a boat enters the water, the hull pushes water away from underneath so that it can float. This is called displacement. From that moment the boat has to work against the friction of the water to move. As with all forms of motive power, it takes more energy to move from a standing start than it does to maintain a steady cruising speed, as a body has to overcome inertia. This is one reason why ships are quite slow to accelerate compared to other modes of transport.

A commonly referred to term in the maritime world is 'knots', which is a measure of a ship's speed. A knot is equivalent to one nautical mile per hour, using the nautical mile which differs from the conventional mile slightly. For example a vessel travelling at 10 nautical miles per hour (10 knots) would be moving at a speed roughly equal to a land speed of 11½mph. Ships have real difficulties sailing smoothly at speeds above 25 knots as they either need to be designed as racing craft, such as a catamaran, that can cut through the water more efficiently, or they need to be lifted out of the water – this is what the hovercraft does and it also achieves another goal in that, unlike its rival the hydrofoil, it is completely amphibious. From the late 1960s and throughout the 1970s hydrofoils were considered by some as a competitor to the hovercraft, but they lacked the manoeuvrability and had to have vast stretches of open water to make them effective. Because of the underwater foils they are also limited in that they have to stay out of shallow waters, especially at low speeds and during berthing. Today the job of the hydrofoil has widely been taken over by the twin hull design of the catamaran.

The hovercraft's skirt, the rubber that surrounds the perimeter of the craft, acts as the suspension for the hovercraft in much the same way as the tyre works for the wheel. There are different types of skirt design, depending upon the size and role of craft. The major two are bag skirts and segmented types. A bag skirt is typically found on large commercial craft such as BHT130 and SRN-4, as a bag is much more efficient and can also be patched if damaged. Segmented skirts are common with small lightweight and recreational hovercraft such as cruisers and racing craft. Here the owner is able to replace individual parts of the skirt quickly and cheaply if they become worn.

The British Hovercraft Corporation (BHC) became the largest manufacturer of hovercraft of the pioneering era, and the Isle of Wight-based firm did much work in the development of skirt design. A hovercraft skirt is one of its most sensitive parts and the design must be just right, or an uncomfortable ride will result. Excessive wear of the skirt can occur if its edges are flapping up and down on the surface of the sea or land. Larger craft, such as the SRN-4 and even today's modern Hoverwork craft, have a bag and finger skirt which enables smooth travel yet allows the ability for different parts of the bottom of the skirt to independently move over obstructions. The fingers act in much the same way as the sections of a caterpillar's body. These fingers also have the added advantage of being easily and relatively cheaply replaced when they become worn.

Some people think that a hovercraft still must be in contact with the ground when they notice the water disturbance behind. This is caused by two things: firstly, if you blow onto a bowl of water then it will cause the water to repel away from the centre of the pressure; what we see at the rear of the hovercraft is exactly that but this is just causing a minimal disturbance to the surface. The other point is that it is only the fingers that seem to skim the surface of the water as they drag along, acting as the crafts' suspension system and helping to retain a smoother and more balanced air cushion. For some the hovercraft's magic was lost when the addition of its skirt became part of its design to enable a more versatile and commercially viable vehicle. The easiest way to tell if a hovercraft is actually appearing to hover is to take a look at front (bow). Here you will see that it is not touching the water and, most importantly, the complete lack of a bow wave as the craft is out of the water. It does not matter what size, shape or role of the hovercraft, its skirt does exactly the same thing and works in the same way. The rubber (or neoprene-coated nylon) is purpose-made, shaped to the craft and fixed to the bottom edges of the plenum chamber slot. As the hovercraft creates lift, the skirt extends below it to retain a much deeper air cushion. The development of the skirt enables a hovercraft to maintain a good operating speed through quite large waves. On larger craft, pilots are able to control the momentum curtain to tilt the hovercraft and assist it to move in a particular direction. This effect is assisted by lifting the skirt at the appropriate point. The action is controlled by the pilot's joystick and the craft will tilt in the direction in which he or she moves it; on a small racing/cruising craft this can be done by the pilot leaning left or right. As we will notice throughout this book, a hovercraft can and will hover without a skirt but having this neoprene-coated nylon cushion creates many advantages, including the reduction of power required to create and maintain a greater amount of lift.

Another idea that seems to baffle people is the question of whether a hovercraft will float, should it lose its air cushion. A hovercraft is designed from the outset to be able to do just that, especially commercial types as they are treated in the same way that conventional watercraft are. Strict safety guidelines have been in place for some time and the answer to the question is yes, a hovercraft will float if it loses power on water. However, it also seems quite strange to think how a hovercraft floats, remembering that the very principle of its operation relies on there being many holes on the bottom of the craft itself! These holes lead to the sealed area of the plenum chamber, the area where vented air is channelled to form the air cushion. It is within this area that most craft hold their buoyancy that keeps the craft afloat. Type of buoyancy varies depending upon the size and type and craft. In small, homebuilt and leisure-type hovercraft, including racing variants, buoyancy tends to consist of expanding foam which is ideal as it is cheap, very light and easy to work with. However, on larger and more industrial craft DIY methods such as foam just wouldn't be up to the job so large buoyancy tanks are fitted in the craft's hull. A hovercraft is therefore safer than a boat but can also be considered safer than an aircraft in the event of a water emergency as they are 500 per cent bouyant.

Creating sufficient amounts of lift itself has always been a major factor within the hovercraft design. A lot of people are also under the illusion that a hovercraft must create a vast amount of lift to enable it to function. While the sight of watching a craft come onto its pad at speed, mixed with a fury of noise and spray, might suggest this, this is not actually the case. Thanks to the job of the plenum chamber a hovercraft only requires a small increase above the atmospheric pressure to create enough lift. In some cases as little as just 4 psi can be measured even on large craft. However, in the early days of hovercraft this was not easy to achieve and a great deal of consideration had to be taken to design smooth air ducts with as few bends as possible.

The next issue is the role of the key apparatus itself, the lift fan. As with all spinning things, the lift fan must be perfectly balanced otherwise it will just shake the whole craft to pieces. Over the years there have been various ways that lift air is created. The main two differences are on small hovercraft where both lift and thrust may be integrated from not only the same power plant but also the same duct. In order for this to work efficiently from a single-engined craft, exiting air must be divided sufficiently between the two. A common ratio on such hovercraft tends to be around 60 per cent thrust and 40 per cent lift. However, on many craft of today, the pilot has the ability to control the amount of lift or thrust air by adjusting the splitter plate which can be found situated within the fan duct itself. This works just like an aircraft's elevator, although again modern sophisticated hovercraft also have a variation of an elevator placed within the exiting airflow behind the duct to alter pitch of the craft when required from time to time to clear obstacles or avoid problems. Some small hovercraft, such as those used for racing, have two engines, one for lift and one for thrust.

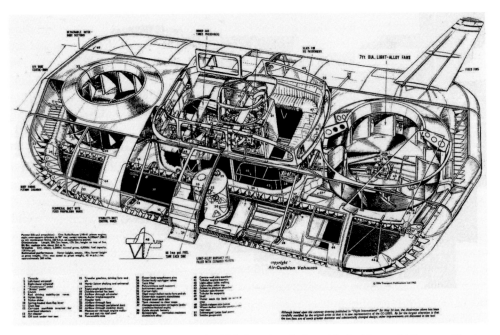

This cutaway image shows how the centrifugal lift fan is installed on this Cushioncraft. The centrifugal duct system is a significant part of the hovercraft, or at least on larger craft.

In most cases, lift fans are inverted and driven from a central shaft to a gear system from its power source. However, there are some hovercraft that have changed this principle by placing two inverted fans at the rear of the craft and making use of this ideal situation by then venting air off for thrust. A British company named Cushioncraft Ltd demonstrated this method of hovercraft design. They quickly became known as the 'whispering hovercraft' due to being much quieter than a conventional design with propulsion ducts or propellers.

Like an aircraft, a hovercraft requires an amount of mathematical calculation to determine its accurate proportions and working out how much power required to create lift is critical. Imagine you had a brief to design and build a new hovercraft that complete with crew, fuel and load weighs 4,000lb and is 20ft long and 10ft wide. If the craft is to hover, the pressure of air creating the cushion must be approximately 4,000lb. This represents 20lb per square foot.

There are many formulas involved when designing a hovercraft but it is quite straightforward to base one on what we have just learned. A cushion of pressure of 20lb can be maintained by 4hp (horsepower) for each square foot of curtain area. Curtain area is the total length of the skirt if taken off the craft and laid on the ground from end to end. A hovercraft of dimensions 20ft by 10ft would have a curtain area of 60ft – twice the length of the craft multiplied by twice the width. If you want the craft to hover 1ft high you would require sufficient power to provide a curtain of sixty times 1 sq. ft. At 4hp per sq. ft you would need 240hp for lift alone.

This gives us a basic understanding as to the main elements that make up a hovercraft and the forces at work that enable it to hover.

THREE

SRN-1: THE FIRST HOVERCRAFT

The world had awoken to the new hovercraft and it was watching developments avidly. Throughout 1959 SRN-1 went through 500 hours of trials that would take it well into the next decade. It was apparent that as the world watched with keen interest a new industry was starting to unfold, and one that would have much potential. There was an exciting buzz around this new hover craze; schoolboys would read about it in their magazines while fathers were interested in building one themselves, saying it was for little Jimmy when the reality was anything but! It was no coincidence that the date for the first cross-Channel voyage by hovercraft fell on 25 July, as this marked the first Channel crossing made in an aeroplane fifty years earlier by French aviator Blériot. A crew was assembled that would pilot the craft across the busiest shipping route in the world. Lieutenant-Commander Peter Lamb had worked with Saunders Roe for many years and had much knowledge and experience in the test flying of new aircraft, in fact at this time he was the firm's chief test pilot. Pilot Bob Stath was intended to travel with the craft but he got left behind at Calais; some say this was to allow Cockerell to travel on his own invention!

The other crew member was the engineer John Chaplin who, like Lamb, had a history with Saunders Roe, being responsible for the firm's fluid dynamics department in control surfaces so his skills would be invaluable in controlling this new method of transport.

So what about the craft itself? Power came from a 450hp Alvis Leonide radial engine which was more commonly connected with aviation. In its inverted role placed within the huge duct the engine would power the centrifugal fan which drew air from the top of its large intake and repelled it at greater pressure on the underside of the craft, whilst a small amount was vented for thrust. These thrusters could be opened and closed to aid in manoeuvrability with rudders placed in the direct exiting airflow.

SRN-1 started its crossing from Calais having being barged safely over where it was prepared and fuelled ready for its maiden flight. The English Channel can at times be unpredictably choppy but the crossing was largely calm. There were, however, a few issues that arose during the journey – these were mostly hazards from crosswinds and swells that affected the overall stability. At the start the craft made slow progress but as more fuel was burned and weight subsequently dropped the speed picked up and it wasn't long before the SRN-1 was skimming the Channel at speeds close to 25 knots. While Peter Lamb and navigator John Chaplin were firmly seated within the doorless cockpit, which wasn't exactly watertight by any stretch of the imagination, the SRN-1 had another character who seemed to be hanging on for dear life on the bow. You might think that this would have been a stowaway or an apprentice from Saunders Roe out to learn the ropes as it were. In fact it was Christopher Cockerell himself, the man at the forefront of the whole

With a gathered crowd from Saunders Roe, SRN-1 enters the water of the Solent for the very first time. Here the craft is being towed while the development team undertake various tests, most notably being flotation in the event of engine failure; as we can see, the craft passed the test.

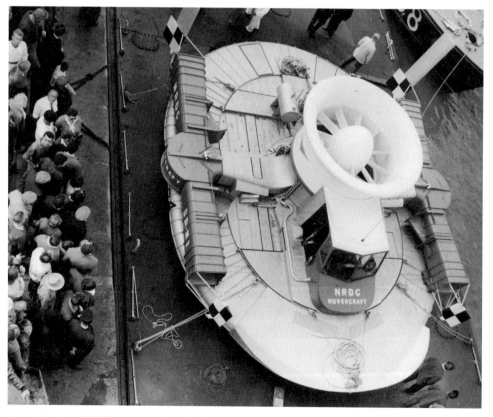

An aerial view of SRN-1 among the onlooking press. From this angle it is quite apparent how the hovercraft was viewed by many as some kind of flying saucer.

story. But he wasn't just coming over for a jolly on a free passage back to Blighty – he had a key role to play, acting as human ballast to counteract the forces of Mother Nature with wind and swell by manoeuvring himself around the craft. The SRN-1 took slightly longer than expected to reach the other side due to weather changes: 2 hours and 15 minutes. By comparison the super SRN-4's last trip was under 25 minutes! But it was still a successful test and by the time the SRN-1 had landed on the beach at Dover, history was firmly made. (It also resulted in partial deafness for its creator, Cockerell.)

In December 1959, the Duke of Edinburgh visited Saunders Roe at Cowes and persuaded the chief test pilot, Commander Peter Lamb, to allow him to take over the SRN-1's controls. He flew SRN-1 so fast that he was asked to slow down a little. On examination of the craft afterwards, it was found that it had been dented in the bow due to excessive speed, damage which was never repaired, and was from then on affectionately referred to as the 'Royal Dent'.

Cockerell thought back to his original early experiments at his boatyard. The sidewall curtain idea made him think further about the addition of a rubber curtain around the side of the SRN-1 within which it would retain more air.

During the constant elevation work it was decided that utilising much-needed air sourced from lift, to then be bypassed for thrust, hampered things and greatly affected the overall performance of such a hovercraft. An alternative power plant was needed to provide a second source of power to generate the much-needed thrust which would leave the large lift fan to do its job of providing greater lift.

SRN-1 on its maiden crossing of the English Channel.

A joyful occasion for all involved in the first hovercraft programme, as SRN-1 glides onto the beach at Dover.

SRN-1 with its later addition of a pointed bow and stern, together with an increased skirt, all of which aided performance and more importantly showed what future hovercraft designs would need to take into account to make it have practical real world possibilities.

SRN-1 is demonstrated to many onlookers in London. It must have been quite a sight to see the craft passing Westminster along the River Thames.

The first hovercraft to appear at the world famous Farnborough Air Show, a showcase for British aeronautics.

The imperial age of steam is rapidly being chased by future promises. Here SRN-1 passes RMS *Queen Mary* in the Solent.

In 1962 a Bristol Siddeley Viper 3 jet engine was added to the rear of SRN-1 to provide the required extra thrust. This separate propulsion engine now accelerated the craft's speed up to 50 knots which was more befitting to its design, concept and purpose. While the noise of an extra power plant was an obvious addition, the overall appearance also seemed to change slightly, with the adoption of a pointed bow and stern.

Hovercraft tend to have a length to breadth ratio of 2:1, unlike boats, being wider with greater beam. Hovercraft don't need to take into account the beam factor, where too much width might make the craft unstable, as they don't float on water like a conventional boat does. The hovercraft uses the cushion of air it creates to float above the surface of the water to ride on. The SRN-1's pointed bow and stern gave an increase in surface cushion area which aided hover height and overall performance of the craft, with longer skirts fitted.

It's not strange to see, then, that by the early 1960s the world had been gripped by this new form of transport that offered many possibilities for the future. Thanks to the men of Saunders Roe, the SRN-1 pioneered the way forward with air cushion vehicles and many lessons were learnt from its trials and development. By now there was increasing commercial, military and recreational interest in hovercraft and it was clear that there was a need for even further research and development.

FOUR

THE NEXT CHAPTER...

The early 1960s was as crucial to the hovercraft as the 1920s was to Hollywood. The 'swinging sixties' saw many new firms spring up all over the world promoting their new designs and hovercraft variants for a wide range of duties. Cockerell no longer had to try to convince anyone that his invention had potential – the world had now realised that.

VICKERS ARMSTRONG

In 1961 one of the most significant aviation companies in the world, Vickers Armstrong, announced a step into hovercraft ventures from their South Marston plant in Swindon, which is now a large car factory. Vickers had a long and successful history with aircraft such as the First World War Vimy, Viscount and VC10 airliners, not to mention their work on Concorde. Vickers also had a testing facility on the south coast by the Itchen River in Southampton. Their first working example of a hovercraft was named the VA.1 which was largely constructed of plywood and glue. Power was even more unsophisticated, dating back to pre-war engines used in Tiger Moth biplanes; a Gypsy Major engine for lift and a Continental C90 for thrust. Although there was nothing wrong with these engines, which were very reliable and sturdy, it just proves the crossover of aircraft components used within developing hovercraft, as nobody had really any reason to think of anything else. However, although primitive in its appearance, the craft proved an important research vehicle for Vickers, teaching the company valuable lessons which would then take them to a more compact craft.

Vickers' next hovercraft set the shape of things to come and offered a layout that remains largely unchanged even in today's modern craft. The VA.2 was a fully enclosed craft but where it differed from most was that it had capacity to carry five seated passengers. It was designed to be air portable and had sides that could be retracted to reduce the width for loading. Powered by three engines, two Rolls-Royce aero piston engines for lift and one 310hp Continental for thrust, the VA.2 had a top speed of 60 knots.

All of Vickers' development work on early hovercraft did not go in vain. Far from it, as in 1962 they produced a new craft that would earn them the title of manufacturers of the world's first commercial passenger hovercraft. The VA.3 went into trial service on 20 July 1962 between the Wirral in Merseyside and Rhyl in North Wales. The planning for this undertaking of a commercial service began the previous year, which shows just how quickly the hovercraft had compelled businessmen to put it into use on a commercial route.

VA.1 on trials in Southampton.

This new 12-ton craft could carry a 2-ton payload and did just that mixing its load of passengers and mail. The commercial operation was under the control of British United Airways run by future airline magnet Freddie Laker. All crossings were listed in their same aeronautical vain as flights. While the previous Vickers craft, the VA.2, may have promised a slightly cleaner appearance, the ugly snipe-nosed VA.3 was anything but. It was a messy-looking craft with four large fins and two engine nasals which spanned equally large propellers. The front fins were later removed which aided pilot visibility greatly. To say that the VA.3 was noisy is an understatement; its four Bristol Turbo gas turbine engines that developed over 400hp each were more used to buzzing over people's heads at 20,000ft and not 4ft as with the case of a hovercraft. As with most experimental aviation cases it would be the engines that would eventually become the craft's downfall. The VA.3 could carry twenty-four passengers facing rear and claimed a top speed of 60mph, which all seemed very promising.

Many early passengers on this pioneer service became familiar with the 'hover-coach', as their new mode of transport to hop over the Wirral had been billed. The hovercraft was scheduled to do twelve trips a day but it only managed six at first, and on four other days only lost two trips due to poor weather – still quite a feat given the new technology involved. In total the VA.3 only ran on nineteen days out of fifty-four. Of the thirty-five days it did not run, the main reasons were strong winds and high seas, and continual failure of the rear lift engine. The last trip was on Friday 14 September 1962. It left Moreton Shore at 1.15p.m. en route to Rhyl. When just over halfway to its destination, one of the lift engines failed, followed in short order by the other. It eventually limped into Rhyl, and although there were two more days of travel scheduled, it was not to be. The next day was very stormy, preventing repairs.

Saunders Roe were not the only firm thinking of large commercial hovercraft for the future. This design concept shows how Vickers proposed a new craft that would fulfil the role that the SRN-4 was to adopt.

An artist's image of the VA.2.

The VA.2 craft was tested worldwide on all terrains.

The prototype VA.2 craft takes shape under the skilful hands of men that have worked on more conventional aircraft models.

Captain Ray Old decided to remain on board overnight because of the winds and in the early hours of Sunday morning the craft broke free from its moorings. Captain Old managed to start the propulsion engine and prevent it drifting out to sea. When the tide receded, the hovercraft was safely beached and remained there till the next day. This time

A period postcard of the VA.3.

VA.3 later in its short life with its forward fins removed, enhanced pilot visibility and improved general dynamics.

all three captains – Captains Old, Colquhoun and White – stayed on board. The winds were even stronger the next day and the craft once again broke free from its moorings, this time drifting out to sea. Captain Colquhoun signalled the coastguard who called out the lifeboat. The craft was by now a quarter of a mile out to sea. The captains boarded the lifeboat and left the craft to the mercy of the wind. It was driven towards the shore and the promenade wall. When it came ashore everyone did their best to secure it to the wall; this they succeeded in doing with great difficulty, considering the pounding it was taking from the waves. Two mobile cranes had to be brought in to lift it out to safety at 7.30a.m. on the morning of Monday 17 September 1962. The craft was eventually blown up and sunk in the Solent in 1971 and is a wreck 70ft under the sea near Ride Pier, and regularly crossed by the Hovertravel ferry still in service today.

SAUNDERS ROE

Whilst Vickers were making noises in the practical use of air cushion vehicles, Saunders Roe were still playing around with the SRN-1 when they announced their next craft. As you would expect it was to be called the SRN-2, but it looked nothing like anything before. This 27-ton craft shared no comparisons to tin cans or MoD-style research vehicles. Instead this new vehicle had clean flowing, aerodynamic lines, a fully enclosed cockpit and cabin and a white paint finish. In fact its design was well ahead of its time; after looking at the rear, note the similarities to a NASA space shuttle! It was so futuristic, it almost looked as though a schoolboy could have dreamed it up and doodled it in his exercise book during the most boring of maths lessons.

But unlike SRN-1, SRN-2 did have a real mainstream purpose: to carry passengers and mail across the busy waters of the Solent. On 11 August 1962 the service began operating between Ryde, on the Isle of Wight, and Eastney in Portsmouth. The operators were a joint force of public transport firm Southdown Motor Services, which were more used to double-decker buses, and Westland Aircraft, who had by this time taken over Saunders Roe. While the VA.3 could carry twenty-four passengers, the SRN-2 could manage thirty-eight plus a crew of two.

The Solent route was a success and laid the foundations for the future of a permanent hovercraft service connecting the mainland to the Isle of Wight. Next on the schedule for SRN-2 in 1963 was a trial service across the Bristol Channel between Weston-super-Mare and Penarth, South Wales, by P. & A. Campbell from 23 July until 30 August.

However, things were moving fast with the hovercraft, and I don't just refer to their motion but more the way in which designers and engineers were pushing the craft further and advancing the concept. Such was the case that after only 610 hours of sterling service the SRN-2, the only one built of its kind, was scrapped and was to be the first ever hovercraft to be broken up. These early craft had four engines for so few passengers so they were never built for profit but to test the concept.

By late 1963 an order was placed with the Ministry of Technology for a new military focused, slightly larger version of the SRN-2 model. The result was, not surprisingly, designated SRN-3 and was launched from the Saunders Roe slipway in December 1963. It was by far the most competent and advanced hovercraft built up to that time, with the ability to carry ninety-two fully equipped soldiers at speeds of over 70 knots. This 37-ton craft was 77ft in length and was the largest hovercraft built at the time. Another unique feature about the craft was that it could also carry vehicles: a development which would turn the hovercraft into the ultimate landing craft. Only one SRN-3 was built and it served under the recently formed Interservice Hovercraft Trials Unit from the Royal Naval base at HMS Daedalus at Lee-on-the-Solent.

A front view of the SRN-2 without its air cushion.

Westland Aircraft were also manufacturers of several helicopter models, used worldwide. It was by no means a radical step for a firm with such an interest in the science of hovering to continue their work in early hovercraft designs.

A promotional sales brochure for the SRN-2 which was used to try to gain commercial interest.

SRN-2 skims the frozen inshore waters of the Solent as it comes ashore with passengers.

The evolution of the hovercraft. SRN-1, SRN-2 and SRN-3 pictured outside the Saunders Roe factory on the Isle of Wight.

By 1974, after using the SRN-3 in a number of roles, the Ministry of Defence decided to use it as a test bed to see how vulnerable hovercraft were to underwater explosions. The craft was tethered and subjected to a series of explosions, even some from explosives placed directly underneath the craft, and yet not only did it survive, but it was still capable of returning to base under its own power. It was broken up after service amounting to 1,430 hours. Both the VA.3 and SRN-3 proved hovercraft to be mine proof.

All these three hovercraft (SRN-1, SRN-2 and SRN-3) were evaluated by the Interservice Hovercraft Trials Unit from September 1966 until December 1974 when the unit was handed over to the Royal Navy. The IHTU continued to put enough stress on SRN-3 to try and sink it, but each time they failed.

★ ★ ★

The site at which the majority of the UK's hovercraft activities were to take place was on the site of HMS Daedalus on the Solent. The site's location was ideal geographically, the situation between Portsmouth and Southampton making it the ideal place, seeing as you could almost see holidaymakers taking their rides to Ryde over on the Isle of Wight – the land of the hovercraft!

Long before the hovercraft unit had set up camp HMS Daedalus had a rich aviation and marine heritage. During the First World War it became apparent that there was a shortage of seaplane pilots that were vital in flying missions hunting German U-boats. There were already training establishments dotted around the country but these were limited and mainly land based. It was decided in the summer of 1917 that they needed a base on the Solent. The strange thing is that it was only meant to be temporary and strict orders were put about that prohibited construction of permanent buildings. Aircraft were to be stored in large tents while all men were in billets nearby. This was presumably for two reasons, firstly to stop the enemy from finding out what the Navy was up to and secondly because it could have been felt that the operation would have been redundant after the war.

It seems fitting that we should learn about the history of the jewel on the Solent that would facilitate the future progression of the new hovercraft.

On 30 July 1917 HM Naval Seaplane Training School, Lee-on-Solent, a unit of the Royal Naval Air Service, was officially opened by Squadron Commander Douglas Evill DSC Royal Navy. The same man would later become deputy to Air Chief Marshal Sir Hugh Dowding during the Battle of Britain in the summer of 1940.

It didn't take long before the station became quite successful; after all, it was the ideal place. The relatively calm sheltered part of the Solent water made it perfect for quite primitive flying machines and equally nervous and unskilled pilots learning to take-off and land on a moving surface. By November 1917 plans had changed from temporary to permanent and large hangars were soon constructed to house the seaplanes. But building wasn't the only change that took place here. 1 April 1918 would be an historic day. The RNAS and the Royal Flying Corps, who had been engaged in their deployment over the skies of the front line, amalgamated to form the Royal Air Force.

By 11 November 1918 the First World War had finally come to an end and facilities were gradually run down. By 1920 the station was busy once again with seaplane activity and also the new HQ for No.10 Group RAF, twenty years later this sector would be responsible for the south-western area of air defence during the Battle of Britain. RAF Lee-on-Solent was a vital airfield for Great Britain, working closely with the Navy to train the best pilots and adapt skills that they would later require for the next war that was to unfold. While the First World War had taught Westminster that aviation was a crucial part in modern warfare, the machines operated were still fragile and easily knocked out the skies, and there was little room for error for the brave young men that flew these planes, often under the fiercest conditions. Things had come on a great deal by the time the Second World War broke out and Daedalus, although not recognised and little known by others, had significance as much as Biggin Hill in the protection of London.

Throughout the Second World War the station underwent further alterations, some resulting from Luftwaffe raids, including concrete runways and formations of new squadrons. In 1943 no fewer than ten front-line fighter squadrons were formed at RAF Lee-on-Solent. Perhaps one of its most prominent operations to take place during the war was its involvement with Operation Overlord. On D-Day itself the airfield became the busiest air station on the south coast of England, principally with gun spotting and fighter reconnaissance. Between 1939 and 1945 eighty-one squadrons operated twenty-one types of aircraft from Lee-on-Solent.

You may also have noticed that during the 1960s many hovercraft sported a BP logo somewhere on their surface. This was because the oil giant recognised that the hovercraft had huge potential for the future and would therefore subsequently become a major customer for their fuels and lubricants worldwide. BP supported the hovercraft in its early days with heavily subsidised or, at least in some cases, free fuel to ensure development projects could continue without having to be jeopardised by running costs. This was quite a positive method of thinking for such a large corporation, considering the hovercraft was still little more than an advancement of back shed Britain technology, compared to the aeronautical industry where BP offered its services to major airports.

When the first cross-Channel hovercraft service opened at Pegwell Bay, BP serviced the needs of operators Hoverlloyd with fifteen purpose-designed Bedford TK Hovercraft refuelling vehicles.

They were clearly proud to support the hovercraft and were quite savvy in their marketing, using every possible method of ensuring the public was aware of the good actions they were doing. BP published numerous brochures and posters promoting the latest hovercraft and future projects. All of this helped to build and sustain the profile of the hovercraft.

While the British Hovercraft Corporation had dominated North Cowes on the Isle of Wight, another quieter part of the island was stirring. Bembridge was a pretty little village which had been the escape of many holidaymakers. The east side of the island was about as far away from the rush of modern life as you can get. It was more about being here and being still than anything else, as you took in the quaint village pubs, landscape and picture postcard scene. However, all of that changed, out of sight and away from public viewing on the local Bembridge airfield, which had itself a rich heritage in the nation's past being one of the most southern RAF Fighter Command airfields during the Battle of Britain of 1940. The airfield had undone countless attacks by German Stuka dive bombers but had remained a private airfield after the war serving light general aviation. On one side of the airfield a small corrugated building-cum-hangar gave rise, quite literally, to an object that would turn Bembridge from being known as a sleepy hamlet of Hampshire to the UK's answer to Area 51.

CUSHIONCRAFT

Britten-Norman Ltd was another British aircraft manufacturer of the time and had a solid reputation for STOL (short take-off and landing) aircraft such as their Islander aircraft which began production in 1965. In 1960 the firm registered a new company as a division to their existing business. Cushioncraft Ltd was born, and Bembridge was to be their original base.

The intention was to design, construct and develop hovercraft and to see what potential they may have as a profitable business. The first contraption appeared not long after SRN-1 had flown across the Channel, when in 1959 Cushioncraft unveiled their CC1. If some felt the SRN-1 was out of this world then they were in for a shock with this one.

It all came about quite bizarrely for Britten-Norman as in 1960 they were approached by the banana distributors Fyffe to construct a new vehicle that could be used as a platform to study the potential of this type of vehicle for the carriage of bananas from plantations in the Southern Cameroons. Together with its associated company, Crop Culture (Aerial) Ltd, Britten-Norman studied the potential for the Cushioncraft in many different countries. These investigations revealed the possibility of a breakthrough in transportation techniques by the use of air cushion vehicles which could accelerate the pace of development in territories where roads are non-existent and costly to build and rivers are seasonally unnavigable.

Cushioncraft's first air cushion vehicle, the CC1. It was a difficult craft to control but proved a very worthwhile exercise for the new company, utilising more commonplace piston engines rather than the costly jet power plants of other manufacturers. Note the wheels.

CC1 demonstrating its hover capability.

The CC1 was their first example and quite an impressive one at that and not just because it was the second-ever hovercraft to rise from planet Earth. It quickly became known as the flying saucer and this skirtless craft had a clear advantage over other experimental designs. For one thing it had an undercarriage with wheels, enabling the pilot to have supposedly greater control when manoeuvring on and off hover. Power came from a relatively small Coventry Climax engine, a variant of a fire pump and later the Lotus sports car engine. Hover height was impressive, though, and ranged from 12–15in. The CC1 never really seemed to do much other than just hover for the press and had a limited usefulness. It was difficult to control at the best of times and on several occasions men with ropes trying to stop the thing from spinning out of control were seen trying to tame it. Lift was generated from a fan the same diameter as the craft itself with exposed workings! Two helicopter tail rotors were used as propulsion, all utilising the same single 160hp engine for both lift and thrust. The CC1 was quite novel and showed that you didn't have to use costly jet engines to produce a working hovercraft. Controlling it, however, would be another issue. We must remember that this was the very dawn of a new type of flying vehicle, and there were inevitable teething problems, much as was the case with the Wright brothers' *Wright Flyer* which could barely manage to turn left and right.

Cushioncraft's next model was totally different and perhaps one of the most attractive hovercraft ever produced. The CC2 looked like it had just hopped back from the twenty-first century. Initially there were no intrusive external engines and even what would seem a lack of propulsion exits. The CC2 used a clever method of obtaining its thrust from using air deflection from the lift engine. This skirtless craft demonstrated its hovering characteristics over a range of surfaces but each time you could clearly note the amazing ground clearance and very noticeable hover. Initially developed to carry eleven passengers and two crew, the CC2 was 30ft long and powered by a single 240hp Rolls-Royce V8 engine, similar to that used in Rolls-Royce cars of the period, but tuned for an increase in power, not that Rolls-Royce would admit to their fine vehicles needing an upgrade for their adequate output!

By now the company had moved to the Duver Works at St Helens where designers and engineers had access to the sheltered waters of Bembridge Harbour. It was intended in 1961 that there would be ten CC2s produced, but in the end only three were built. In 1962 CC2 models 001 and 002 were sold to the Ministry of Technology, who then extensively trialled them from their new base at the Royal Aircraft Establishment in Bedford. This airfield was also home to the two largest buildings in Great Britain, the old airship sheds that once housed the famous R101. RAE Bedford shared some of its goings on with its sister site at Farnborough which around this period was developing the strategic TSR-2 supersonic aircraft, a jet so advanced that even today its systems cannot be matched by serving aircraft. This was another example of the rapid advance of, and greatly developed interest in, technology during this troubled era of political uncertainty.

Whilst under their new ownership of government interests the CC2 underwent radical changes. To begin with, the most notable was the addition of two external propulsion engines which increased the operational speed from 45mph to 50mph. A segmented skirt was also added which increased hover height and resulted in more capable craft for its intended military applications.

A Cushioncraft CC2, under the control of the Army, in a clear state of hover under trials.

The Royal Aircraft Establishment also trialled the CC2 and this craft is fitted with external propulsion engines to enhance its performance. The addition of the RAF insignia on its tail shows that the hovercraft was being seriously considered by the Air Force, despite the knockbacks Christopher Cockerell had in the early days of trying to win the backing of government support.

A CC2 craft skims its way into the water for trials.

The CC2 was perhaps one of most glamorous hovercraft ever built, making it even more of a shame that none of them survived.

Humour and novelty has always been the route of the hovercraft and this photograph highlights the case as a CC2 hovers in to fill up. While we can imagine Rovers and Austins in the line, it is most likely that this pump was actually used for aircraft at Bembridge.

Schoolboys look on in amazement at the CC2 hovering. It had a 12in ground clearance and a speed in excess of 40mph.

CC2 hovering in a field.

The CC2, which had by now been fitted with a segmented skirt along with its external propulsion engines, comes ashore at speed.

A Cushioncraft CC7 undergoes stability control tests with obstacles placed in its path. Note that as the hovercraft negotiates a steep bank, air pressure is still retained via its skirt. The CC7 was quite a transition in hovercraft development of the era, featuring inflatable side decks, and was the first Cushioncraft model to use a gas turbine power plant.

HOVERTRAIN

While Britain was getting excited about hovercraft, the rest of the world was also paying close attention. In France a coach was adapted with an air cushion platform to reduce friction. It was fitted with very narrow tyres which reduced friction further and also aided steering. But road vehicles were not the only means of transport being adapted by the French; trains were too. This was, of course, still very much the jet age and from 1965 a bizarre-looking polished aluminium hovertrain appeared.

Jean Bertin, a forward-thinking French aeronautical engineer, had visions of creating a sophisticated high-speed rail service that would bring people and cities closer together. In 1956 Bertin formed a new company and started work on his elaborate and much-needed plans for modernising the rail network of his beloved country. Bertin used an idea of the air cushion, an idea that was rapidly sweeping the world, and he knew that it had definite possibilities for what would become known as the Aérotrain project. The first French Aérotrain, model number 01, was a half-scale experimental vehicle weighing 2.6 tons. It was originally propelled by a three-bladed reversible-pitch propeller powered by a 260hp aircraft engine, which was later replaced by a small aircraft jet engine. But Aérotrain couldn't run on existing rails, so a derelict line was chosen in Gometz-le-Châtel in Essonne which would be adapted into a 6.7km test track for the project.

Although a train does not need to have the amphibious qualities to that of a hovercraft, the Aérotrain still made good use of a form of air cushion in order to function. Having

The French-built Aérotrain glides across the countryside on its purpose-built experimental track, some of which still survives today.

The Aérotrain under construction, utilising much aviation skills in its creation.

slippery steel rails for trains is ideal to reduce friction of rolling wheels but what if you could raise the train from the rail altogether and do without the wheel? It sounds quite bizarre but that is exactly what the Aérotrain does, an idea which has to this day, with a few modifications, been carried forward into twenty-first-century railway systems, particular those in Japan. Aérotrain 01 used two 50hp compressors to generate the lift under its body.

Much experience and design work was carried over to the next Aérotrain concept, which again like its predecessor was a half-scale example. Aérotrain 02 used a JT12 turboprop engine from Pratt & Whitney, and in many ways was more a wingless aeroplane than train. It looked like a cross between a lavish speedboat, rocket, aircraft and spaceship. It must have been a sight to see one of these things running alongside technology from the previous century; even the odd steam locomotive was still in sterling service on the railway lines of France!

But real progress was made with Aérotrain S44, the first to look like something that could have commercial potential and one that would no doubt make admiring schoolchildren long for a high-speed ride. It was hoped that Aérotrain S44 could be used for inter-city commuter routes at speeds of up to 200km/h. The vehicle used a linear induction motor, a form of magnet propulsion which is a clean and relatively quiet and effective system, although a downside is that they can require vast amounts of power to function.

Quite possibly the most elegant of the Aérotrains, model 180 was a fully functioning vehicle that could have changed the face of the French high-speed rail network. It was over 75ft long and weighed over 11 tons. Most significantly, it could carry up to eighty passengers in air linear refinement. A proposed top speed of 250km/h seemed feasible with very limited friction and plenty of propulsion power coming from its two Turbomeca Turmo gas turbine engines, each delivering 260hp. Another unique feature of this Aérotrain was its ducted propeller which helped to augment its pleasing appearance, not to mention its aerodynamics. A further gas turbine engine was installed which ran twelve compressors, six to provide the air cushion and six for guidance on a vertical plane. However, as with all hovercraft, going is not the problem, stopping is! Aérotrain 180 made use of reverse

A concept model of the Aérotrain, showing its airliner-style seating configuration.

A six-passenger Aérotrain runs along an inverted T-section rail.

thrust to slow and stop it, but under emergencies a friction brake would be applied onto the single rail which it used for guidance.

From 1965 until 1974 five prototype Aérotrains were built. On 5 March 1974 Aérotrain I80 gained the world speed record for an overland air cushion vehicle, with a top speed of 267.4mph; quite a feat for a train of the period. Without a doubt Jean Bertin's Aérotrains paved the way for the backbone of establishing a high-speed rail network through mainland France and today's TGV trains owe a lot to the men that worked on the Aérotrains. Thankfully some of the Aérotrains have been preserved, along with a section of the original test track in Essonne.

In Britain things were slightly different. By the late 1960s and early '70s lots of hovertrain concepts were flying around, which mostly consisted of monorail systems. One such example was a fully working prototype using a magnetic levitation (maglev) system, a very simple approach but one that does require a lot of electrical power to maintain sufficient energy in the electromagnetic workings of opposite poles repelling each other in order to create the desired lift. These are still hovering trains though!

This idea was shelved due to the rising costs and the lack of available technology at the time, as the computer systems had not been invented to run them efficiently. By 1972 the British were experimenting with another type of train, one that despite the oil crises had the support of British Rail. In the post-war age of switching over to diesel power, it was clear that the British railway system needed to be updated to maintain competition with motorways. A high-speed line was needed and furthermore a high-speed train that could transport its passengers in comfort between major cities at high speed. The problem was that Britain was just too small to have a network like the TGV uses in France, with its long, flat, straight lines where trains can maintain high speeds. The answer was a revolutionary and complex new approach to the train itself, for it to tilt into the bends like a motorbike. This method would mean that the high-speed train could run at great speeds without the need to slow down too much for corners. The APT-E, Advanced Passenger Train Experimental, was the first incarnation of this method, and used a gas turbine engine to provide its power. It never entered commercial service, being used to test the principle. It was axed in 1985.

A vision for the future – how the hovertrain was viewed as the railway in the sky.

If anything at all cries 'weird vehicle', then this has to be it. The second experimental Aérotrain to be
built looked like a cross between a Second World War fighter and something from a 1950s Dan Dare
comic, with its polished aluminium bodywork.

SIDEWALL HOVERCRAFT

Another interesting version of the hovercraft to appear around this time was the sidewall hovercraft, which only required lifting the bow and stern of the hull out of the water while the sides would remain in contact with the water. The idea behind this came from making the skirt almost rigid by using a conventional boat hull design, for the side parts at least. Its principle was often referred to as the 'air riding catamaran' and is the nearest missing link that separates boat from hovercraft in the history of the air cushion vehicle. Sidewall hovercraft are also called Surface Effect Ships (SES).

In 1968 the first commercially successful sidewall hovercraft entered service. This was the HM.2, manufactured by Hovermarine at Southampton. The main advantage with this design is that the sidewalls trap the air along the length of the craft thereby increasing lifting efficiency, while noise and spray is greatly reduced. Like on a catamaran, the twin sidewalls increase stability and control, while propulsion is also aided by water screws exiting from the sidewalls from a common form of marine diesel engines. At low speeds they are economical and easier to control within the tight confines of locations such as river moorings. However, there is one drawback with the sidewall hovercraft and that is that they are limited to sea state as they are not amphibious. In the early 1960s sidewall manufacturing was at first only carried out by Denny, the shipbuilders from Dumbarton, Scotland, but the company went into liquidation and subsequently it would be Hovermarine that would make the sidewall into a global product. Denny did, however, produce a workable craft, the D3, which commenced a hover-bus service along the River Thames in London and even seaborne ventures, proving that the sidewall concept had definite possibilities.

Hovermarine was formed in 1965 and quickly wasted no time in creating their first sidewall design with the HM.2. Manufactured not from aircraft aluminium or marine steel but GRP fibreglass, this new craft could transport sixty passengers in comfort and had a lower length to beam ratio than its predecessor the D2, with a cruising speed of 35 knots. The first HM.2 craft was operated by Seaspeed in 1968 between Ryde Pier and Portsmouth Harbour.

After various redevelopment work, orders came in from all corners of the world. One craft even went to operate on the world's highest navigable lake, Lake Titicaca in Bolivia. Alas the air is thinner, so at altitude lift is less, and the craft was not as efficient. However, the most significant order came from Hong Kong in 1974. The Hongkong and Yaumati Ferry Company (HYF), at the time the world's largest ferry operator at sea level, saw the new HM.2 as having all the qualities they required in a vessel: it was fast, efficient and offered smooth travel for its commuter and pleasure passengers. In total the HYF had thirty HM.2s in operation. At one point the HM.2 was an equally familiar sight in Hong Kong as the red Routemaster bus was in London for thirty years.

The early HM.2 suffered from mechanical problems and during a financial crisis in 1969, Hovermarine went into voluntary liquidation and the major assets were acquired by a new company, Hovermarine Transport Ltd. This company was subsequently taken over by the American company Transportation Technology Inc. Production of HM.2s for North and Central America was then undertaken by Hovermarine's factory at Titusville, Florida. Over 110 HM.2's were manufactured in the UK and US. But though the British were the first to develop the sidewall craft successfully, in the United States the US Navy had been investigating into the role an SES might play within its fleet.

In the 1960s the USA was thinking hard about what a surface effect ship could do to aid its naval operations and concluded on the construction of two 100-ton experimental craft. Speeds of over 80 knots were regularly attained under trials, giving a high-speed option for various roles. The second craft, 100B, was built by Bell Aerospace (Textron Corporation) at the NASA Michoud assembly plant. Propulsion was provided by two driven, semi-submerged, reversible propellers, which allowed this craft to achieve the almost insane speed of over 96 knots (approximately 110mph!) whilst under testing conditions in the Gulf of Mexico near Panama City, Florida.

Not content with simply having the attribute of speed, the craft also demonstrated its fire capabilities by launching the Navy's first vertically launched missile whilst doing 60 knots. This showed how stable the design of the craft's hull was and also that it didn't have to compromise over speed while carrying out practical roles.

The 100A and 100B were both capable of operating almost completely out of the water on an air cushion with only 18in of sidewall, the propellers or the waterjet inlets entering the water. The SES programme showed the US Navy that a sidewall hovercraft had plenty of scope to enter an operational role within the forces. However, it had its drawbacks, as it did not meet the requirements for a vessel that could enter a combat zone at high speed, deploy its needed cargo of a tank, troops and equipment then turn back out to sea at speed, a role that was and still is fulfilled by the LCAC hovercraft.

During the 1960s inventors tried to utilise hover technology in almost any field; it had been fantasized about for decades and now there actually seemed a way that childhood dreams of travelling to school in your dad's hovercar could become real. Visionaries went to bed thinking about hovering. In fact, even beds received the hover treatment: the world's first 'hoverbed' was designed by English designer Lesley Hopkins for use in hospitals. These were especially beneficial for treating major burns victims, where gently levitating them away from the bed prevented their wounds from sticking to sheets, while the cool circulating air gently healed and soothed the skin itself. The idea was ingenious, although quite bizarre. Hoverbeds are still in use today in special wards and their design continues to assist in the medical treatment of those with major burns. There is almost one in every hospital in the United States, although only five in the whole of the UK.

Then came the hovercar. I will discuss this subject further in a more relevant chapter, but I will explain about the hovercar craze that swept the United States in the early 1960s. It was more than just a revival of the '50s cartoons that promised flight for all vehicles of automobile size. Most of the big motor manufacturers were producing elaborate and sophisticated designs; the '59 fin tail Cadillac was perhaps one of the most iconic of the period for just being so extreme, ostentatious and beautiful, yet unreliable. The American Dream, the ethos of the United States, was about prosperity and success. In the definition of the American Dream by James Truslow Adams in 1931, 'life should be better and richer and fuller for everyone, with opportunity for each according to ability or achievement,' regardless of social class or circumstances of birth. The idea of the American Dream is rooted in the United States Declaration of Independence which proclaims that 'all men are created equal' and that they are 'endowed by their Creator with certain inalienable Rights' including 'Life, Liberty and the pursuit of Happiness.'

It is this very selection of words, this term that has forever stuck within the loyal hearts of the American people with their hard work, determination and positive attitude that they can achieve great things for themselves as an individual and for a sustainable family life. During the late 1950s anything seemed possible and, after all, why shouldn't it? The nation had embraced and was living a time of colour, glamour and excitement. The sci-fi age had begun and was being absorbed by the mainstream. Drive-in movies screened the latest space-age thrillers as young lovers cuddled in the sunset, sat within the comfort and security of their open-top Buick Roadmasters.

Flying car concepts were being pushed out of Detroit quicker than Henry Ford could paint a Tin Lizzie black and motorists wanted more. Although most of these hovering offerings never made it past the stage of glamour models draped over the bonnets and into the open skies, they did suggest there was a viable market for such a vehicle which could set pulses and purses racing across the nation. But how could it work? News reporters would talk about the latest General Motors idea of a flying car but nothing was actually appearing in the market place. Would the hovercar ever take off? Or would it just live its days from motor show to *Life* magazine feature?

Marilyn Monroe may have personified the optimism of Hollywood, but the V8 was the spirit of the automobile world. Highways were getting busier and busier but the skies seemed full of promises. All over the country enthusiasts, engineers, designers and just about anyone who fancied doing something about it themselves experimented with making cars fly. Unfortunately there were casualties but there were also some very interesting things happening. If the automobile industry was dragging its heels with the hovercar idea then the aerospace companies were not.

AVROCAR

In 1958 Avro Aircraft Ltd of Canada, a division of the British aviation company responsible for the wartime RAF Lancaster bomber and Cold War-era Vulcan bomber, came up with a version of a hovercar called the Avrocar. It looked more like a flying saucer than anything you'd see outside a diner and was capable of vertical take-off and landing (VTOL). However, the Avro car was not one that was ever intended for the average man on the street. It was the result of a top-secret defence project for the US military to explore the potential of the flying disc concept which is known as the Coanda Effect, which is a form of jet named after its designer, Romanian aerodynamics pioneer Henri Coanda, who was the first to recognise the practical application of the phenomenon in aircraft development. The Coanda Effect works by forcing pressurised air out of the end of a tube, and then over the top of a metal disc. The Coanda Effect makes the air stick to the disc, bending down at the edges to flow vertically. This airflow supports the disc in the air.

At this time in our history it was expected that any future European conflict, which being at the height of the Cold War was eminently possible, would start with a nuclear exchange that would destroy most airbases, so therefore aircraft would need to operate from very small makeshift locations, roads or even unprepared fields. A great deal of research was put into various solutions aimed at a second-strike capability. Some of these solutions included rocket-launched aircraft, while many companies started to work on VTOL aircraft as a more appropriate long-term solution.

The Avrocar project was led by its designer Jack Frost, who had worked at British Aircraft manufacturer de Havilland and was responsible for some of Great Britain's important aircraft of the era, including the Hornet and the Vampire jet fighter. This decade of aviation development saw engineers the world over trying hard to develop supersonic aircraft that could have practical 'real world' capabilities and Frost was no exception, having been the chief designer at de Havilland on supersonic research. To fly faster than the speed of sound (approximately 760mph at sea level) was what every aviation firm and indeed air force wanted for the future. Frost looked into Frank Whittle's design of the reverse flow engine but he felt that he could do something much more simple and yet more effective. Frost built his own engine which he referred to as the pancake engine.

The Coanda Effect plays an important role with aircraft aerodynamics. Air passing over the aerofoil section of a wing can be bent down towards the ground using the flaps which aircraft use to slow themselves down; a form of air brake. This bending of the airflow increases speed and decreases pressure where the lift is increased. Another form of aerodynamic activity is also taking place under this effect, on the boundary layer. In the case of an aircraft, the boundary layer is a thin sheet of air which attaches to the immediate surface, almost as if the wing creates an invisible skin as it travels through the air and along the ground at speed. The boundary layer is one of the most important parts of airflow and its merits are also used within the design of motor vehicles. The understanding of these forms of aerodynamics is crucial in order to think about unorthodox flying and hovering vehicles.

John Frost of Avro Canada spent considerable time and effort researching the Coanda Effect, leading to a series of strange-looking, inside-out, hovercraft-like aircraft where exiting air passes in a ring around the outside of the craft and could then be directed by a series of flaps in this exiting circular flow. This is as opposed to a traditional hovercraft design, in which the air is blown into a central area, the plenum, and directed down with the added use of a fabric skirt.

There is no question that in the air the Avrocar would have resembled a flying saucer. Two prototypes were built to prove to the men in black that the concept was a viable one. Remember that this was all some time before the Harrier Jump Jet had appeared at the Farnborough Air Show and the very idea of a vertical take-off and landing aircraft, other than that of a helicopter, was nothing more than a commercial aeronautical firm's pet project. Proposals were put through for a possible USAF fighter and a US Army reconnaissance aircraft. However, even after the Avrocar eventually made it off the ground and away from its publicised state of hover, controlling the craft was no easy task Its thrust problems and stability issues meant that it could never operate under the conditions that was hoped. Throughout the programme, the Avrocar was referred to under different names, from Omega, Project Y-2 and Spade, and for a further model of the type, Silver Bug, all of which may reflect a top-secret operation not far removed from the realms of a Hollywood sci-fi movie. The Avrocar project finally came an end in 1961. Somewhere within a highly secret R&D establishment on US soil an even more deniable operation was taking place, culminating in their experimental work of achieving what would have been felt even by the world's aero firms as the impossible, let alone by the public: a supersonic, high-altitude flying disc.

The Avrocar's designer, Frost, worked hard on a range of various propellers and estimates for a craft that could travel at Mach 3.5 at 100,000ft. While much work was invested in these strange flying discs by both the US Government and private aviation firms such as Avro, other activities were taking shape which required the very in-demand funding, such as the race to beat the Soviets to the Moon. Only now when we look back at these decades of the late '40s to the late '60s can we understand how and why the pace for aerospace development had advanced so rapidly, as well as its importance to mankind itself.

The physics that make a hovercraft function can therefore be linked to the goings on of the many UFO sightings, and even NASA.

At some time there were, excuse the pun, high hopes for the Avrocar and its siblings. The US Army even used Avrocars depicted as flying jeeps in company literature; a role that the Huey helicopter would later fulfil.

There was a second Avrocar which managed around 75 hours of flying time. It too was considered by some as a failure due to its performance and overall achievements. Despite the sums of money invested in the programme, what had the US Government got? An aircraft that couldn't fly? Or a highly probable new vehicle that could be used to assist the nation in its Cold War defence capabilities? The fact is that, despite its failings, the Avrocar kept the Soviets guessing. Maybe they thought the whole thing was little more than a smokescreen for something much more sinister the Americans were developing, something that even today we could only dream of. Avrocar was just a rubber skirt shy of being one of the world's first hovercraft.

If the Avrocar had ever reached the public domain then the world would by now be a very different place. What the Avrocar did do is inspire, most notably the work on the Moller skycar, which was also of circular disc design but utilised ducted fans instead of costly and noisy jet engines.

However, as with all forms of transport the flying car, hover car or future car has always had one major drawback to becoming a reality: legislation, the dreaded red tape of bureaucracy. This is not necessarily a bad thing; as much as I hate the little things that get in the way of enjoyment and the unnecessary filling in of forms, turning a car into an aircraft is a very grey and dangerous area. How does a driver, who will no doubt have a sense of urgency to get into his or her flying machine and take to the skies, qualify overnight to become a pilot? The problem is you can't, and a flying car is no longer a car when it is in a state of flight; its weight is supported by the air itself and in terms of control systems it becomes an aircraft. Controlling such a vehicle would mean having the necessary skills and understanding that conventional pilots share. All this comes before even considering the numerous other factors at work, including navigation, maintenance and weather!

Imagine stepping outside your house, kissing goodbye to your wife, husband, dog or all of the above and stepping into your Ashley Hovercar XP001. You switch on the radio and listen to a fine choice of music as you start the engine and glide out of your drive and onto the road. The sun is shining as you lift straight off into the clouds to join a skyway network with a pre-programmed route via the on-board GPS. But as the vehicle is travelling through the clouds the weather starts to change, the sky appears grey and your vehicle won't climb past the weather. You can't climb into the stratosphere as the craft is just not designed to go that far, and if you don't run out of oxygen then you'll most probably freeze to the controls at -30°C, perhaps before the whole thing breaks up.

Pilots need to be trained in meteorology to understand how the weather affects all aspects of flying, their routes and how they can avoid getting into danger. All pilots study the local weather in their pre-flight planning no matter how big or small the aircraft or route. It just wouldn't have been realistic for the Average Joe to do, at least not during the decades of advancement and uncertainty.

A strange combination of events that seemed to occur around the same time with the sci-fi age of the '50s and the ever-increasing sightings of UFOs in the '60s. Not long after the famous Roswell crash of 1947 people had started to give thought to the night sky and strange goings on. I am not a ufologist, I am not a government press agent, employed to discredit alien sightings, and I haven't been abducted by ET either. What I am is a person who has done a lot of travelling and seen a great deal, most of which can be explained, but some of which maybe cannot. As we have seen with the Avrocar, the height of the Cold War dramatically increased the productivity of research and development departments in defence.

The ability to hover brings so many advances to a military machine; to be able to sweep down upon an enemy, gather information, launch an attack or just see what they are up to attracts the interest of government officials. Looking at all the various information reports of UFO sightings, mainly concentrated throughout the United States during the 1960s, it is apparent that something quite unusual was occurring in the skies. Sightings of strange spinning discs that would show up, seemingly travelling very slowly through the sky and in a lot of cases fixed with a state of hover before zooming off at impossible speeds. The silver discs, which would quickly become referred to as 'Flying Saucers', seemed to share an appearance similar to that of the Avrocar. There is a lot of unknown myth around this period in aviation development but it is also reasonably well known that there were designs, concepts and indeed aircraft that were so far ahead of most people's thinking that if they would have seen a new secret project then they could quite conceivably have thought it would have come from another world.

During the Second World War a number of rushed aircraft designs were tested based around the idea of using circular wings. While the Allies were busy in America, the Germans were being just as secretive, if not more so. In the secluded mountains and valleys of the Alps, the secret division within the Third Reich had constructed a labyrinth of underground tunnels carved deep with the mountains and well out of sight of inquisitive eyes. What were they doing in there? Whatever it was they didn't want the advancing Allies to capture their secrets. Towards the end of the conflict the Nazis withdrew from the complex site but not without destroying as much of it as they could.

It is rumoured that these tunnels hide the ingenious scientists that created much of this top secret work, their bodies entombed in the rock forever along with the project they were working on within the shadows. Despite the size and scale of Nazi engineering and special projects operations, and the time of the demise of the organisation, very little can be proved as to what they were up to. We can go on accounts from eyewitnesses, blurred photographs and myth. What is clear, though, is that a selection of German scientists were captured from the site at the end of war by the US Army and were taken back to the United States where they were housed within a secure air base in the Nevada Desert. These men were no ordinary boffins, they were rocket engineers, and had developed the supersonic V2 weapon. It is common knowledge today that the Apollo programme had German design directors from the V2.

Some time after the Second World War, rumours surfaced that the Nazis had had one more secret weapon up their sleeve that was still hidden away from Allied eyes: an alleged disc-shaped aircraft. According to these stories, some of the Allied nations had plundered the German laboratories where these aircraft were being developed and secret testing of these devices explained many of the reports of flying saucers that appeared in the United States and the Soviet Union in the 1950s, and possibly even the unexplained sightings over Washington DC. It is quite magical to think that while the rest of the world was trying to avoid the horrors of the Second World War a small team of Nazi boffins were working away in a highly guarded and secured facility in the mountains. What they were alleged to be working on was regarded as the cream of the Nazis' 'Wonder Weapons', a flying saucer.

There are accounts that state German aircraft designers were working on several disc-shaped aircraft towards the end of the war. At a facility near Breslau, Poland, a close team of highly skilled engineers constructed a prototype circular air vehicle of over 137ft in diameter with a pronounced hump on top for the cockpit. This craft was to be powered by adjustable jet engines. It is said that the device was destroyed when the plant where it was being constructed was blown up by retreating German troops before it could be overrun by the Soviets in 1945. But having heard tales of flying saucers in the USA, what if this could have got into the 'wrong' hands – those of the Americans?

Viktor Schauberger, an Austrian Nazi boffin, was considered by many as quite an 'off the wall' thinker. Viktor did not start his career as a scientist, however, but as a forester, which helps to explain some of the ideas behind his creations; several of his inventions used vortex-like patterns that replicated nature.

Like so many inventors, Viktor was not of a highly noble academic background and yet what he did was nothing short of pure genius. He had a deep knowledge of biology, physics and chemistry. Schauberger believed that machines could be designed better so that they would work with the flow of nature rather than against it.

One of Schauberger's projects was to produce a flying machine, saucer shaped, that used a vortex propulsion system. His theory was that 'if water or air is rotated into a twisting form of oscillation, known as a "colloidal", a build-up of energy results, which, with immense power, can cause levitation.'

After the war had finished Viktor emigrated to the United States as he had been informed that the government, possibly the CIA, would fully support and fund his ideas further. However, none of this actually came to pass. Instead the opposite happened and Viktor signed all of the rights over to the relevant departments within the US Government, most likely the Air Force, and was promptly sent back to Austria with his tail between his legs. The misery hit him hard, everything he had worked for, his own designs and thoughts of the anti-gravity solution had been given away at no cost and with no creditation. It didn't take Viktor long before he would give up on himself and he died only five days later, on 25 September 1958, a broken man.

Schauberger had written to a friend that a full-sized prototype of one of his designs was constructed using prison labour at the Mauthausen concentration camp. It is reported that this craft flew on 19 February 1945 near Prague and obtained an altitude of 45,000ft in only 3 minutes! The letter goes on to say the prototype was destroyed by the Nazis before it could be captured by the Allies.

The low stall/drag of the disc shape would have been particularly interesting to the Germans at the end of the war. Months of bombing had reduced German runways to rubble. A saucer-shaped craft might have been able to lift off the ground with a short runway or even do a vertical take-off and landing with no runway at all.

Sadly there is little concrete evidence, other than what is now second hand, that German flying saucers were flying during the time of the war. It certainly would seem plausible that if they were developing this new 'wonder weapon' for whatever reason then they would want to keep it under wraps, even if that meant destroying the device itself and maybe the men that created it. But what would be the point? The Allies would now be in control of Germany, the Nazi party abolished, so who would benefit from keeping it quiet even when they have been defeated?

There are so many theories on this and it's not for me to delve into them here but one thing is for sure and that is that there were a huge number of exceptionally intelligent people working on hidden rocket and advanced aircraft designs at the time of sightings of disc-shaped craft by Allied pilots flying over Europe. Secret the plans might have been, but what was common knowledge was that the United States and the Soviet Union wasted no time in developing rocket technology immediately after the war, and the knowledge must have come from somewhere…

THE 'HOVER ROVER'

In Britain things were much more down to earth and liable to be discussed over a nice cup of tea with a scone, thank you very much. Land Rover, not the first name that springs to mind when thinking about space-age vehicles, had also placed a mark with the hover age. This was the work of Vickers Armstrong, another aircraft manufacturer that had also constructed hovercraft to their own designs.

The Hover Rover of 1962 was never meant to be a vehicle to lust over but was aimed to be a practical solution to the agricultural issue of crop spraying. Technically an air cushion craft, the Hover Rover was a converted Series II 109 pickup, and was developed to mini-mise ground pressure, allowing the vehicle to traverse boggy terrain. The main problem was that it required two engines, vastly reducing the payload capacity. Although the hover cushion supported much of this extra weight, it was still a problem for conventional road operation (although the skirt could be raised for road use). The Hover Rover was never developed further, although for a few years Vickers used it as a promotional device for their larger hovercraft. It proved a novel approach to another application in which the hover-craft could help but unfortunately it never made it past the prototype stage – three were made but converted back to cars.

Towards the end of the 1960s and into the early '70s the hover car idea was little more than a memory of a fantasy time. There were those that would continue their dreams and aspirations in amateur-built concepts and even companies that would go on into forth-coming decades reasserting and heavily investing in the future of such sky cars, but all of this has as yet come to little more than just experimental vehicles. It was a fascinating time to live in: engineering and science were taking new shapes and forming the future of a new beginning, leading to what we have today. Hidden research work will no doubt con-tinue and come up with further advances in time. Today the car manufacturers are looking back at the ideas and have artist impressions of what could be!

Land Rover's version of the hovercar, a mix of the future and the 'go anywhere' 4x4 vehicle.

The Vickers Armstrong firm had hopes that their Hover Rover would be an answer to the agricultural issue of crop spraying and other such roles. However the project was relatively short-lived.

The Hover Rover turned out to be no match for the conventional 4x4, despite what this image may lead you to think.

Vickers were always keen to demonstrate the advantages of the craft as here among a pretty rural and once peaceful setting!

The Hover Rovers were acquired by Land Rover in chassis cab form and when the development and experimental work was completed, they were all sent back to Land Rover where they were overhauled and re-bodied, to find themselves in showrooms as new vehicles. Perhaps some of these vehicles exist today as pickup trucks and station wagons?

FIVE

THE BRITISH HOVERCRAFT CORPORATION AND THE SRN-5/6: HOVERING THE WORLD

Saunders Roe laid a path towards generating their most iconic model. Up until this point the hovercraft for them was all about development and they needed a product they could sell. That was not a problem, as they had interest from all over the world, and what they eventually put forward ticked everyone's amphibious vehicle boxes. On 11 April 1964 the new SRN-5 hovercraft from BHC was launched from Cowes and marked a turning point in hovercraft history. This new Warden class hovercraft mixed everything they had learnt over the years and used a new Rolls-Royce gas turbine engine, a variant used in the Folland Gnat Red Arrows fighter jet of the period. This new hovercraft had a dumpy appearance and in its full hover had a wave clearance of 5ft.

On first sea trials it was found that it was difficult to steer the hovercraft out of any prevailing winds due to the large tail fin. This was later rectified with a dorsal type fin and the rear surfaces were reduced in size. Throughout its trials model 001 went through various modifications, including venting air from the rear of the craft which was diverted to movable ducts to assist the craft manoeuvring at low speeds and in restrictive conditions. 001 was seen in the same waters as the operating SRN-2 on the Solent passenger route, so it was no surprise that this new craft would join in the commuter route, for a short period at least, from 17 June to 21 August 1964.

The hovercraft then surprised crowds at the Farnborough Air Show of that year, making high-speed runs along the runway. But the show didn't stop there; next it was pressed into service alongside the IHTU SRN-3 in Operation Armada. At last, a hovercraft that could be put into mass production.

During the height of SRN-5/6 production the British Hovercraft Corporation bought into Cushioncraft Ltd in 1967 to invest in a minority share holding, and it revived the name under which Britten-Norman's initial ACV endeavours were launched. Cushioncraft Ltd was then entirely bought out by BHC after they had financial problems in 1971.

Early operations of gas-turbine-powered hovercraft highlighted the problems of running this type of engine in a salty and dusty environment. The original configuration of the SRN-5 had a pipe running from the eye of the lift fan to the engine compartment. A mesh filter unit was then added on a fairing on the top of cabin ahead of the engine, but this could still be subject to salt spray. Trials with plenum intakes started on 6 January 1966. These intakes drew air from the plenum chamber under the craft, and passed it through a series of filters.

SRN-5 – the world's first hovercraft – on the production line.

Assembly of the main cab. Fifteen SRN-5s were built.

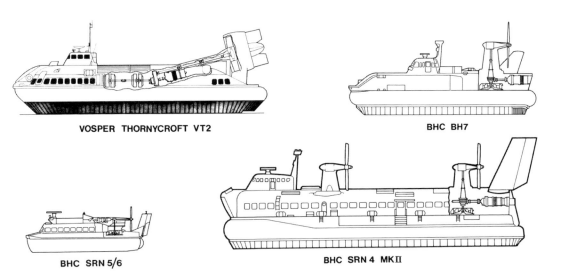

VOSPER THORNYCROFT VT2

BHC BH7

BHC SRN 5/6

BHC SRN 4 MK II

Size comparisons of four utility craft used by the military.

The prototype SRN-5 001 leaving the factory in the waters of the Solent. One of these craft has been preserved by the Hovercraft Museum Trust.

The craft had started its life with a jetted skirt, but advances in skirt technology were to be tested from February 1966 when a fingered skirt was fitted to the craft. As previously mentioned skirts vary with hovercraft design and the subsequent years of development. In 1971 SRN.5 001 was acquired by Air Vehicles Ltd. This company converted the craft into a flat-deck freighter, and on 31 May 1972 the craft had its first flight after the conversion. On 30 June 1972 the craft was moved to the Wash. The craft returned to the Isle of Wight on 26 August 1972. From 25 May 1973 until 28 September this craft was operated with SRN-6 022 on a service from Southport to Blackpool for Hoverwork.

In November 1974 001 was shipped to Australia. In February 1975 it commenced operations in the Torres Strait for the Department of Aboriginal Affairs. Then in 1977 the craft was lying at the Hoverwork yard at St Helens on the Isle of Wight, having been put up for sale. In 1979 the craft was broken up, after a total of 3,148 running hours. It still holds the record crossing of the Solent of 4 minutes 53 seconds, although this was unloaded.

The Warden class SRN-5 was 38ft long and could carry eighteen passengers including the two crew members. Entry was via a lift-up door on the bow which offered ample head room and inside the cabin was laid out in a cross between an RAF transport aircraft and ferry. This was because from the outset the SRN-5 was designed to be versatile, to be able to adapt to a variety of roles including carrying passengers, cargo or even ammunition for the battlefield. Some pilots found that the craft was almost too fast for its own good due to the power of its 900shp engine that drove a variable pitch, Dowty four-blade propeller. The SRN-5 went through a hard series of trials with the newly formed IHTU at HMS Daedalus in Lee-on-Solent across from the Isle of Wight.

The unit operated SRN-5s in military service whilst still under trials; in the Far East they saw action in Borneo and in Malaysia during the Indonesian confrontation. They were particularly successful in night patrols against insurgents and for amphibious jobs in hostile conditions such as swamps and over rapids, clocking up over 1,000 hours in the field.

Overseas orders for the SRN-5 were initially strong and it was to be the Americans that placed the largest single order of that time. Six craft were supplied to Bell Aircraft, a company that had a background in hovering as they had developed another larger craft, the 'Hydro Skimmer', and of course various helicopters. The crafts were adapted with a separated off cabin area and an engine that they could easily maintain in the field; for simplicity it had to be a power plant they were familiar with and trusted, so American General Electric engines were fitted. Bell then adopted a new name for their crafts – Bell SK-5 (SK standing for Sikorsky, Bell's founding company). Other modifications included US military radar, various shielding and a .50-calibre machine-gun mount. Throughout their time in Asia the hovercraft would see a variety of roles with different armaments. The famous river boat patrols that have since been immortalised in many Vietnam war movies also tried the hovercraft, or PACVs (Patrol, Air Cushion Vehicle), which were commonly referred to among GIs as 'Pak-Vs'.

The Vietcong could no longer rely on evasive tactics used to defend themselves from passing Huey helicopters, and now they faced much faster patrols as the hovercraft could reach in 1 hour a target it would take a gunboat half a day to reach. Also, having fallen foul of the popular misconception, they thought that all they had to do to disable a Pak-V would be to puncture its skirt. They were proven wrong. The hovercraft was a force not to be argued with, though two were destroyed in action.

While they were faster they also had some disadvantages in this area. Firstly, you can't creep up on the enemy in a hovercraft as they make a lot of noise. Secondly, they were hugely

SRN-5 004 and 017 in service with the US Coastguard as SK-5.

Into battle. During the Vietnam conflict the SRN-5 was rebuilt by Bell Aerospace in the US for use as a versatile assault and patrol craft. Here an example is seen operating as part of the famous river patrols.

expensive to operate and maintain in comparison with a conventional boat. In fact they cost about the same as a Saber fighter jet and at close to $1 million each it wasn't hard to see why this was too expensive. Using aviation technology in a situation with limited and mainly automotive technical skill meant they were often quite complex to maintain in action.

In January 1967 they went for general overhaul and re-equipment back to the USA, before three US-built craft returned in 1968 for a further year in Nang and Tan My.

After Vietnam three SK-5s were taken to San Francisco where they were used by the Coastguard, no doubt accomplishing many rescues which would have taken longer with other boats.

The SRN-5 design was later stretched to offer more seating capacity and greater load space. This simple transformation, into what became the SRN-6, paved the way for the most successful early production hovercraft and helped BHC stay afloat (pardon the pun) during what could have been a tricky financial period. An impressive total of eighty-six SRN-5 and SRN-6 craft were built.

The larger sibling of the SRN-5 was the longer SRN-6 Winchester class, which used the same engine and components. From the outset this was designed as a fast ferry, accommodating thirty-eight passengers or 3 tons of freight. The SRN-6 quickly became the craft of choice for the newly established hovercraft operators for commercial routes: Hovertravel and Seaspeed on the Solent, Hoverlloyd on the Channel, and Hoverwork, a unique charter service.

The SRN-6 was made all the more famous when one example was taken on a 2,000-mile expedition up the Amazon and Orinoco rivers in South America, during which the craft negotiated up rapids and violent rocks that not even canoeists or any power boat would dare to brave. This could not have been achieved without the skilled pilots who learned their trade travelling through the busy Solent. Hoverwork, Hovertravel's subsidiary company, took two SRN-6s to Canada in 1967 and ran a passenger service for an event, Expo 67. Canada had always been a favourable operating region for the hovercraft with its wide, open plains and rough northern terrain. Canada quickly saw the potential and ordered the SRN-6 for search and rescue and was still in use until 2003.

The longest hovercraft journey of the time was achieved by an SRN-6. This was the British Trans-African Hovercraft Expedition, which covered over 5,000 miles through eleven African countries between October 1969 and January 1970. These amazing journeys also spurred off the Matchbox models that small children could now play with.

By the late '60s the SRN-6 was *the* hovercraft and these new operators learned a lot from what would set out their future business structures. In rescue cases a hovercraft could offer a similar but more extensive capability than that of the helicopter but with a much reduced cost and a much greater payload.

The British Forces extensively trialled and used the SRN-6 in a variety of roles, most remotely with the Royal Marines in the Falkland Islands in the South Atlantic. They could carry fifty-five fully equipped troops or 6 tons of equipment. However, the hovercraft were not in the Falklands when needed. When the war broke out between the UK and Argentina over ownership of the rocky islands the hovercraft unit had been disbanded. Prime Minister Margaret Thatcher even made enquiries to try and borrow Saddam Hussein's new hovercraft just finished at Cowes!

The SRN-6 came in many variants, these including the single propeller and dual propeller variants, and both were used around the world for different purposes, such as by the Canadian and Saudi Coastguard. A freighted version was also produced, consisting of a flat-decked standard SRN-6 and fast attack patrol craft for Iraq. Initially developed, trialled and manufactured from 1963 by Westland in parallel to the SRN-5, the SRN-6 carried over 500,000 passengers annually from the mainland to the Isle of Wight until newer diesel craft came into operation in 1982.

A well deck SRN-6 Mark 5 craft of the Royal Marines lies ready for action. They could carry a range of cargo from Land Rovers to pallets and soldiers.

A well deck craft on trials in freezing conditions demonstrates the harsh environments in which the hovercraft can operate.

A SRN-6 craft of the Royal Navy being unloaded back in England following an exercise to the Falkland Islands.

The crew of SRN-6 XV859 inspect the beached remains of a whale during a visit to the Falkland Islands.

The SRN-6 twin prop lies parked on the beach at the Hovercraft Museum at HMS Daedalus, Hampshire. This important craft has thankfully been saved by the museum.

SRN-6 of Hovertravel comes ashore with fare-paying passengers. Hovertravel invested heavily in the SRN-6 craft and it proved a great success for the new company.

SRN-6 009 skims along at speed on an expedition which was provided with fuel by BP.

Due to its unique nature, the hovercraft had great appeal to those that sought marketing and promotional ideas. Here a canvassing Liberal politician is forced to get his feet wet when the SRN-6 he had chartered for a campaign to take him around the south coast of England went unexpectedly wrong when the craft lost its power and had to be towed, or in this case pulled, onto land. Such breakdowns, though, were quite rare.

SIX

BHC SRN-4 MOUNTBATTEN CLASS: THE 'SUPER 4'

When most of us think of the hovercraft we might remember a personal sighting and for many members of the British public that would be the huge cross-Channel car ferry SRN-4s that operated from Dover to Calais. Simply, this was the hovercraft of all hovercraft and the largest in the world.

By the mid-1960s the hovercraft had evolved rapidly in a short time. By comparison, the list below highlights the development of five forms of transport:

	Demonstration	First Service	Gap of
Automobile	1885	1901	16 years
Aircraft	1903	1919	16 years
Helicopter	1907	1950	43 years
Hydrofoil	1919	1952	33 years
Hovercraft	1959	1962	3 years

This gives you some idea of the pace at which this relatively new and fresh method of transport was springboarded into public service. Early trials with SRN-2 and the VA.3 had proved potential but in both cases no certainty that running such a service profitably over the long term was possible. SRN-6 was active on a new commercial route with the even newer Hovertravel, who dared to pioneer a full, commercial hover service linking the Isle of Wight with the mainland in 1965.

Early in 1966 Westland and Vickers hovercraft divisions amalgamated to form a new company that would now become the world's largest and most experienced hovercraft manufacturer. The result was the British Hovercraft Corporation (BHC) which would become solely owned by Westland.

Swedish American Line and Swedish Lloyd, two historic Swedish shipping companies, had for many years tried to enter the English Channel ferry market without success. P&O and British Rail 'Sealink' (British Rail's shipping division) successfully blocked them by not allowing their berths in Dover and Ramsgate to be used by anyone else.

It was the coming of the hovercraft that encouraged these two companies to look seriously again at entering the Channel ferry market; after all, hovercraft do not need berths, just a beach. Investigations quickly showed that BHC had on the drawing board the SRN-4 car and passenger-carrying hovercraft but at this time, 1966, its completion was a good year away. BHC said they would carry out route-proving trials with the smaller SRN-6 Mark 1, and it was this that prompted the two Swedish companies to join forces and form Hoverlloyd.

With every form of transport there comes a time when it reaches a state of supersize. The Rolls-Royce is a large car, a super tanker is a massive ship and it is a wonder the Airbus A380 ever gets itself off the ground. The hovercraft is not left out and on 26 October 1967

An impressive spectacle. The huge SRN-4 passes under Tower Bridge as it hovers down the River Thames in London for a marketing event.

One of the SRN-4s preserved at the Hovercraft Museum today. (Author's collection)

the largest hovercraft ever built glided out of its factory in Cowes on the Isle of Wight. The SRN-4 had been in planning for a while but the success of the smaller SRN-6 allowed research and development to continue.

This was a worrying time for ferry and air travel operators with services across the Channel as this new mammoth hovercraft would be set to attract custom, although quite how much was unsure in the early stages. All this in only ten years from Cockerell's coffee tin experiment!

This new craft was like nothing else before. It had four of the world's biggest ever variable pitch propellers, at 19ft in diameter, mounted on four equally towering pylons that would swirl to make the craft extremely manoeuvrable. It was 130ft long with a 78ft beam. Not only could it take 254 passengers, it could also carry over thirty cars! Weighing in at 160 tons it was powered by four Rolls-Royce Marine Proteus turbine engines with a maximum output of 4,250shp each, up to a maximum speed of 75 knots. There were also two smaller gas turbine engines installed that would power the control systems and provide all of the auxiliary power.

Each Proteus engine was connected to one of four identical propeller/fan units, two forward and two aft. The propulsion propellers, made by Hawker Siddeley, were of the four-bladed, variable and reversible pitch type. The lift fans, made by BHC, were of the twelve-bladed centrifugal type, 11ft 6in in diameter. Since the gear ratios between the engine, fan and propeller was fixed, the power distribution could be altered by varying the propeller pitch, and hence changing the speed of the system, which accordingly altered the power absorbed by the fixed pitch fan. The power absorbed by the fan could be varied from slow (i.e. boating), no lift generated, up to its maximum, 1,000 turns a minute, with full lift generated. The craft could cross the English Channel between Dover and Boulogne in about 35 minutes, taking almost an hour off the time taken by conventional sea ferries.

In 1970 Hoverlloyd announced that they were looking at the possibility of a service to Ostend, and on 7 October, Hoverlloyd's *Sure* operated a charter flight from London (Tower Pier) to Tilbury. It was announced that if the third London airport was built at Foulness,

SRN-4 free-flight model tested and shown at Osborne, Isle of Wight.

A Rolls-Royce Silver Cloud exits a Rolls-Royce-powered hovercraft.

The engines of SRN-4 are carefully installed. The craft uses four Rolls-Royce Marine Proteus turbine engines, the same power plant that Concorde used!

in the Thames estuary, Hoverlloyd would investigate providing a service into London. During the winter period, 1 October 1970 to 1 June 1971, Hoverlloyd announced a 65 per cent increase in passengers carried.

Hoverlloyd operated many return flights daily from Pegwell Bay to Calais.

In an incident in late November 1971, Hoverlloyd's *Swift* was stranded at Sangatte, near Calais, with most of the skirt ripped away. *Sure* was on overhaul and so service had to be suspended. Despite this incident the public were not deterred and business was proceeding very well but Hoverlloyd needed extra capacity, so in late 1971 they ordered their third SRN-4 Mark 1. This was handed over on 14 June 1972 at a cost of £1.75 million. GH-2008 was named *Sir Christopher* (after its inventor) on 29 June and it made its maiden commercial flight on 3 July.

By the end of 1972 Hoverlloyd were still short of capacity and took the decision to send all three craft back to BHC for conversion to Mark 2. The overall size of the craft would remain the same but the internal layout and cabin width would be changed. The first craft, *Swift*, was flown back to Cowes in September 1973 with the other two craft following throughout 1974; *Swift* was back in service in mid-January 1974.

The service was proving popular: between the start in 1969 and the end in 1974, Hoverlloyd carried 3,715,000 passengers and 540,000 cars. But trouble was looming at Swedish American Line and Swedish Lloyd. Increased costs in the 1970s forced the company to look at re-flagging to a different nation, but negotiations with the Swedish Unions failed, and the last two conventional ships were sold in 1975 which meant Hoverlloyd's profits were siphoned off back to the parent company, and at this point the two compa-

Two SRN-4s in production in 1968.

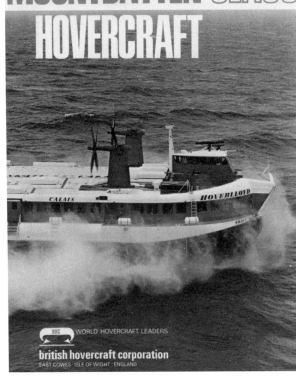

A promotional brochure for the
Mountbatten class SRN-4.

Even a craft of this size must be tested to make sure it floats. Here the first SRN-4 undertakes buoyancy evaluation in the Solent. Each and every craft was provided with inflatable lifeboats for all passengers and crew, just the same as a conventional marine vessel.

nies became known as the 'Brostrom Group', named after the initial chairman of Swedish America who was Dan Brostrom.

Nevertheless Hoverlloyd had 30 per cent of the 'short crossing' English Channel traffic, and other profitable services, such as chartering a craft to the British Ministry of Defence (MoD). Between 4 and 6 May 1976, Hoverlloyd's GH-2008 *Sir Christopher* visited Portland for mine countermeasures trials for the MoD, calling at HMS Daedalus (now home to the Hovercraft Museum) at Gosport, Hampshire, en route.

Despite reports that the company was not making any profit and rumours that Hoverlloyd and Seaspeed were to merge, Hoverlloyd ordered a fourth craft, *The Prince of Wales*, which was the first SRN-4 to be built to Mark 2 specification from scratch. *The Prince of Wales* (GH-2054) entered service between Ramsgate and Calais on 18 June 1977, the craft having more than 140 modifications and improvements on the original SRN-4 Mark 1, the most obvious difference being the longer, new type of control cabin.

In early 1978 Hoverlloyd announced that they were considering the purchase of N-500 Hovercraft from the French company SEDAM or possibly stretching its SRN-4 craft. But in 1979 talks of a merger between Hoverlloyd and Seaspeed were again resurrected as Hoverlloyd's owners announced in September that its four hovercraft were for sale. The British Railways board described the talk of merger as 'pure speculation'. In 1981 the Hoverlloyd and Seaspeed merger was again discussed as £1 million per year could be saved by setting up a joint engineering and maintenance base, and without a merger, all cross-Channel hovercraft operations were in danger of closing down. The Monopolies

Commission reported that it could see no realistic alternative to the closure of British Rail Hovercraft if losses on the scale of those recorded in the past continued. Hoverlloyd's owners, Bronstroms Rederi AB, announced that if a merger did not take place then they would close their operations as soon as possible. Although Hoverlloyd made money for its owners, Bronstroms faced financial problems from other subsidiary companies and Seaspeed made hefty losses over the period 1977–80. Seaspeed's *Princess Margaret* was off service after hitting the Prince of Wales Pier in thick fog on 23 January. With *Princess Anne* refitting, Seaspeed chartered the *Swift* and then the *Sir Christopher* from Hoverlloyd before *Princess Anne* returned on 23 February. The go ahead for the merger of Hoverlloyd and Seaspeed was given during the summer and Hoverspeed was formed in October 1981. Hoverspeed was officially 'launched' on 25 October and until March 1982 six daily return flights were made each to both Calais and Boulogne from Dover. *Sir Christopher* was the first craft painted in the new red, white and blue livery. 1982 saw Hoverspeed carrying 2.5 million passengers and 400,000 cars in the first six months – the same number as the two companies had carried in 1981, but with 35 per cent fewer flights and with 250 staff having been made redundant. By the year's end, the company had a 21 per cent market share of cross-Channel traffic. Plans were announced to extend the former Hoverlloyd SRN-4 Mark 2 to Mark 3 configuration. On 31 August the *Swift* received a tear in its skirt and beached nearly a mile west of Calais. Passengers and cars were offloaded on to the beach where they were picked up by the *Sure*, demonstrating the versatility of the craft. Ramsgate Hoverport was closed to cross-Channel traffic in September and became the Hoverspeed maintenance base.

PRINCESS MARGARET

(SRN-4 GH-2006 001. Mk 1 & Mk 3)

The first Mountbatten class SRN-4 was launched at East Cowes at 2.07p.m. on Sunday 4 February 1968. The craft cost £1.75 million and was fitted with a 2.5m-high Mark 1 skirt, which was expected to cope with most conditions in the Channel. It underwent 2 hours 30 minutes of trials, covering a distance of 20 miles and reaching speeds approaching 50 knots – this in winds gusting to force 6. The world's first hovercraft car ferry made its maiden flight from Dover to Boulogne on 11 June, crossing in 35 minutes.

Further test runs were undertaken before the Mountbatten class SRN-4 returned to East Cowes for final completion. The outward trip from East Cowes to Dover was made in 2 hours at an average of 56 knots. The prototype SRN-4, now named *Princess Margaret*, entered commercial service for Seaspeed on the 26-mile route between Dover (Eastern Docks) and Boulogne at 10.25a.m. on 1 August 1968, following two days of press and VIP trips. This route was chosen so that customers could easily be switched to British Rail's ship ferry service if anything went wrong. The hovercraft was officially named by HRH Princess Margaret on the previous day; both she and her husband, Lord Snowdon, crossed to Boulogne and back. Three days later the craft was out of service with a damaged skirt and a small oil leak. It returned to service on 8 August, although the two-month trial period was marked by a lack of reliability. From 15 August to 30 September 1968 six daily return flights were advertised, the first leaving Dover at 08.20a.m. and then every 2 hours.

Hovercraft have an exceptional safety record but there have been one or two accidents. On this occasion when entering Dover port in high winds *Princess Margaret* was blown into the sea wall.

Wednesdays were a half day — with three round trips followed by a period of maintenance. Fares were £3 10s (£3.50) single with a day excursion for £3. Even the moderate
wave conditions to which the Board of Trade's initial licence limited the SRN-4 played
havoc with the 2.5m skirt. It emerged later that a chain linking the inner flaps of the segmented skirt for extra strength was not up to the actual stresses received and when it broke,
the loose ends slashed the skirt fabric to ribbons. Changing skirt sections was a lengthy
operation and Seaspeed only had maintenance facilities at Dover and no craft in reserve.
During November, *Princess Margaret* was withdrawn and returned to Cowes for the fitting of the new Mark 2 skirt which had been developed, as well as some fairly substantial
alterations in certain other specifications.

Princess Margaret returned to BHC in the summer of 1976 for stretching, 55ft being
added amidships, making her the first Mark 3 craft and the largest hovercraft in the
world (being given the pet name 'Super 4'). On 23 January 1981 *Princess Margaret* was
off service after hitting the Prince of Wales Pier in thick fog; the craft was flown back to
Cowes for repairs. On 30 March 1985 *Princess Margaret* had her darkest day when she hit
the southern breakwater at Dover. The craft sustained severe damage, with four passengers losing their lives. The craft itself was too severely damaged to fly back to BHC and
BHC staff were taken to Dover for a six-week period to carry out repairs. Following
the inquiry, which found the captain was at fault, he was dismissed never to pilot a hovercraft again.

Princess Margaret was retired to the Hovercraft Museum on Lee-on-Solent and arrived
on 16 December 2000. In total she had completed 48,195 hours of operation up to
1 October 2000, and even had the honour of appearing in the James Bond film *Diamonds
Are Forever* in 1971.

SWIFT

(SRN-4 GH-2004 002. Mk 1 & Mk 2)

On 10 December 1969, Hoverlloyd's first SRN-4 (production number 002), later named
Swift, was rolled out on to the pad at East Cowes. It was the first craft with the new Mark
2 skirt which provided both a smoother ride and more protection to the bow. Hoverlloyd's
SRN-4 underwent trials at the Pegwell Bay terminal on 17 January 1970. On 23 January
1970, the craft was named *Swift* by Mrs Mary Wilson, wife of Prime Minister Harold
Wilson. Hoverlloyd's new Pegwell Bay service commenced with the SRN-4 *Swift* on
2 April. Exactly one month later, the Duke of Edinburgh officially opened the £1 million
terminal and flew *Swift* across the Goodwin Sands himself.

In September 1973 *Swift* was returned to BHC where it was widened to increase
capacity from 250 to 276 passengers and from thirty cars to thirty-six, making it the first
Mark 2 craft. *Swift* completed 22,419 operating hours up to 29 September 1991, before in
1994 being donated to the Hovercraft Museum, who kept it for ten years before letting
Hoverspeed cannibalise it for spare parts to keep the Dover service running! Vital parts of
Swift including the cockpit were kept. Like the *Princess Margaret*, *Swift* was also a star of the
screen, featuring in an episode of *The Professionals*.

Hoverlloyd operated four SRN-4s and two SRN-6s.

Hoverlloyd ran from 1966 to 1981, Ramsgate to Calais being the first international service.

SURE

(SRN-4 003 GH-2005. Mk 1 & Mk 2)

Hoverlloyd's second SRN-4, *Sure*, was named by Mrs Soames, wife of the British Ambassador to Paris, on 3 June 1969. *Sure* returned to BHC in early 1974 for the same treatment as *Swift*. Hoverlloyd broke up *Sure* for spares in 1983, a decision they later regretted when they found themselves short of capacity. *Swift* and *Sure* were so named to give a swift and sure service. *Sure* had completed 17,852 operating hours up to 1983. They were the most used craft in the early days! Real workhorses.

PRINCESS ANNE

(SRN-4 004 GH-2007. Mk 1 & Mk 3)

The second Seaspeed SRN-4, *Princess Anne*, started operations at Dover on 8 August 1969. The craft was officially named by HRH Princess Anne at Dover on 21 October 1969. The twin Dover craft (*Princess Anne* and *Princess Margaret*) together achieved a 95 per cent reliability rate.

Princess Anne returned to BHC in the summer of 1977 for stretching, 55ft being added amidships, making it the second Mark 3 craft. As with *Princess Margaret*, this craft was also taken out of service on 1 October 2000 and was flown to the Hovercraft Museum on 3 December 2000. *Princess Anne* had completed 46,418 operating hours up to 1 October 2000.

Princess Anne set a new cross-Channel official record of just 22 minutes on its 10.00a.m. flight of 14 September 1995. Its Master, Captain Nick Dunn, said, 'The conditions were just right, calm seas, excellent visibility and not much traffic in the shipping lanes. I just opened up the throttles and the craft's four Rolls-Royce turbines did the rest.' The craft

Princess Anne and her cargo of light hovercraft embark upon a visit with members of the Hovercraft Club of Great Britain to Calais.

All sorts of people used the cross-Channel service, from holidaymakers to pleasure riders and business users. These cars exit and board the waiting *Princess Anne*.

had undergone some preparation for the attempt and the load that day was light. Nick Dunn told me that the run was made on the return flight to Dover, after having parked *Princess Anne* facing the sea at Calais. All that had to be done was to lift off and push the throttles to full power, heading in a straight line for Dover. I was told that the usual speed limits within the harbour at Dover were waived on that occasion, and the craft was brought towards the pad at full power. As the tide was out the slope of the ramp to the pad was used to brake the craft together with a generous amount of reverse pitch on the propellers. The craft was then unceremoniously dumped on the pad to make the record time.

SIR CHRISTOPHER

(SRN-4 005 GH-2008. Mk 1 & Mk 2)

Sir Christopher made its maiden flight on 29 June 1972 and underwent the same conversion to Mark 2 as *Swift* and *Sure* in early 1974. *Sir Christopher* was the last craft to leave Ramsgate following the merger of the two companies; this was on 12 July 1987. Up until that point it had been held in reserve for two years at Ramsgate. *Sir Christopher* completed 19,116 operating hours up to 29 September 1991, making it the most used standard SRN-4. The craft was finally scrapped for spares in 1998, and like *Swift* kept *Princess Anne* and *Princess Margaret* in spares!

PRINCE OF WALES

(SRN-4 006 GH-2054. Mk 2)

The *Prince of Wales* made its first flight on 3 July 1977 and was the first craft to be built to Mark 2 specification (rather than converted). It operated until 1993 when it was gutted by fire while undergoing maintenance, by which time Hoverlloyd had merged with Seaspeed to form Hoverspeed. The *Prince of Wales* had completed just 11,755 operating hours up to 29 September 1991.

* * *

In total six SRN-4s were built yet for a time the military were considering placing an order as their sheer size meant that they could fly straight over a landmine unscathed whilst carrying vehicles, troops and cargo. In the event of war the MoD would plan to commandeer the craft and in 1973 they conducted trials with SRN-4, *Sir Christopher*.

Here's a comparison for you. A modern airliner of the late '60s, the Bristol Britannia, used the same engines as the SRN-4. While the Britannia would fly in the stratosphere at around 400mph, the SRN-4 would manage 65mph and fly only 9ft or so above the choppy waves of the English Channel. However, the SRN-4 was only 6ft longer than the 100-passenger-capacity Britannia, which could carry 258 passengers and thirty-six cars, not to mention the odd coach. When you view it like that you have to appreciate that a hovercraft of this scale is actually quite capable. In thirty-two years of service the fleet of six had completed a quarter of a million operating hours, while carrying 75 million passengers and 11 million cars.

While the SRN-4 was in service another rival design entered the Dover–Calais route. The SEDAM N-500, launched in 1976, was a radical design; it differed in many ways from the SRN-4 but was just as much a giant. It should have carried 385 passengers and forty-five cars at 76 knots (87mph) but this vision was never achieved. This was the result of the work carried out by engineer Jean Bertin, who had been the lead engineer on the Aérotrain, and the French Government who liked the idea of their own hovercraft to ferry passengers back and forth across the Channel. The most advanced feature on the N-500 was its skirt, which was made up not of the conventional bag type that surrounds the perimeter of the craft, but a series of very large pressure cells which made the large craft look like it rested upon a set of enormous upside down plant pots. This design ensured that the French did not have to pay a penny to the UK for using the patent.

The skirt system developed by SEDAM and Bertin consisted of forty-eight independent skirts, with half arranged around the periphery and the rest forming an inner envelope. The two volumes thus contained were at different pressures. The primary advantage of this solution lay in the independence of the skirts; each could 'react' to local obstacles independently, and each could be controlled separately through regulation of the airflow feeding it. But it was an over-complicated system which brought up the old question, why reinvent the wheel?

The craft was also much more space aged in its appearance, with a large winged control surface at the rear where three powerful 3200 Lycoming gas turbine engines propelled the 240-ton machine. There were further two sole lift engines giving a 6ft clearance on

The French-built N-500. It was hugely costly and very unreliable. While the craft had some interesting features it proved no match for the SRN-4.

Here you can see the unique skirt system of the N-500 which featured a strange air cushion provided by a series of circular segments which looked like inverted flowerpots.

the air cushion. Unlike the SRN-4, the N-500 had an upper deck for its passengers above the car deck. It joined the Seaspeed fleet in 1978 but, despite its similar capacity to the SRN-4 craft, was withdrawn out of service by 1981, due to being much more costly to run and always out of service! Hopes were initially high for Naviplane, a hovercraft with French-designed skirts, though only two craft were actually built. The first was destroyed after a major fire during maintenance, having only completed 7 hours of service, while the other craft developed major technical and structural problems, and was later handed over to the scrap man in 1985. The N-500 was totally unreliable, but remains another important chapter in hovercraft development: it had proved interest from another nation apart from the UK, and explored some radical thoughts into further improving the concept for the benefit of its users. It could, however, be argued that it would have been so easily improved using British know-how – the patents the French wouldn't use. I for one think that it was quite a remarkable thing and one that its creator, Jean Bertin, would have been proud to have seen, had he have lived to witness its operation.

The SRN-4 had a flight crew of three which consisted of Captain, First Officer and a Navigator, much the same as aircraft of the period, including Concorde. But despite the similar crew, hovercraft could carry far more people. By far the biggest alteration of significance to the SRN-4s occurred in 1976 when two were stretched, adding 55ft amidships, making them 185ft long and with a new raised skirt which provided an increase in hover height to 12ft. The result was that the new Super 4s ran at a lower pressure, giving a better ride. Passenger capacity was now increased to 423 with sixty-one cars on the deck. These two craft, *Princess Margaret* and *Princess Anne*, now had a total gross weight of 320 tons. But it clearly wasn't to come cheap and at £12 million each for the refit it caused quite a stir! The local press thought that the increase would bring further costs to the customers and many wondered if the company would ever recoup their investment. These new Mark 3 Super 4s meant the cross-Channel service could operate even in 57mph winds or a force 8/9 gale, with 11½ft waves, making their entire operation that much more amenable. Alas BHC made a loss on the stretch!

In 1981 the two competing SRN-4 companies merged together to join forces as Hoverspeed. At this time competition was high and newspaper campaigns were offering discount tickets for their readers to travel to France. However, people didn't take the hovercraft because it was cheap, they took it because they wanted to reach their destination as quickly as possible and also because they wanted to feel the adventure of flying, the excitement of the entire hovercraft experience. Loading and unloading was so much faster than the ferry as you didn't have to walk miles to your car and once in your car there was sufficient room to just drive straight off without having to wait for a berth.

Hoverspeed continued to operate their seven SRN-4s and one N-500. Over the history of its creation the hovercraft has one of the best safety records of any vehicle in the world and is by far one of the safest forms of transport, due to their buoyancy and skirt cushion despite the speeds!

Sadly many good things in life come to an end and in October 2000 the SRN-4 hovercraft service was discontinued. Many people think that the reason why the hovercraft stopped service to Calais was due to the soaring costs of fuel and the Channel Tunnel becoming more popular. But this is wrong. In fact the service ran for ten years against the Channel Tunnel and was still faster! The final nail in the coffin was the fact that the EU stopped duty-free shopping in 1999. This meant the end of the situation where passengers could buy their goods tax free aboard the craft. Newspapers would offer constant promotions where readers would travel to France on the hovercraft for just £1, a promotion that was worthwhile due to the prospect of duty-free sales on board. It was hard-hitting and the operation would cease very quickly.

Hoverspeed totally ceased trading in 2005 but they left behind an amazing and iconic part of our history. Today the last remaining craft, *Princess Margaret* and *Princess Anne*, are resting at the Hovercraft Museum in Lee-on-Solent. Visitors can still see the huge presence that these machines offered and thankfully they have been preserved for future generations to be amazed.

After much careful research I have added an account from former BHC employee and hovercraft enthusiast Mr Bob Hanna, who recounts below the last flights of the SRN-4s. This book is about preserving the facts, heritage and wonderful achievement of the hovercraft. The SRN-4 represents, just like the more famous Concorde, a part of our national heritage. It is so much more than just an instrument of transport and belongs to our nation. It is with thanks to the volunteers and trustees of the Hovercraft Museum that we can preserve such machines for future generations. Our flying boats and our liners have gone, and like Concorde one should be preserved. I could not leave this out of this book as I feel strongly that it sums up the admiration of us all and just how special the SRN-4 Mountbatten service was.

FLYING SUPER 4

By Bob Hanna

Princess Margaret, GH-2006
3 September 2000

A little under three weeks before its ultimate retirement from service, I had the honour to join the crew of SRN-4 Mark 3 GH-2006 *Princess Margaret* for a series of flights between Dover and Calais. My host for the afternoon is Capt. Linton Heatley, Hoverspeed's projects captain. Linton hails from New Zealand and joined Hoverlloyd as a seasonal first officer (FO) in 1973. In addition to being a Master Mariner, he also holds an ATPL (Airline Transport Pilot's Licence), and has professionally flown fixed-wing aeroplanes before moving to hovercraft full time. Since 1973, he has amassed some 11,000 hours on SRN-4s, including 5,500 as pilot in command.

Assisting Capt. Heatley are FO Peter Bird, also a SRN-4 qualified pilot, and Second Officer Roger Warren. The FO on a SR-N4 operates as flight navigator, while the SO occupies the co-pilot's seat, operating the engine and systems controls, and is referred to as the flight engineer (FE). Thus, the natural progression of promotion has been from right seat to navigator's panel, thence to left seat. It is Hoverspeed policy that crews should all be able to operate one position below their normal role. Today, Roger, a flight navigator and a captain himself before semi-retirement, is taking the right seat. The flights are also crewed by a CSD (Cabin Services Director) with eight cabin attendants, plus a car deck supervisor and his crew of five.

Whilst very hazy, the winds are light and the Channel is flat calm. Linton directs me to climb a steep ladder in the centre of the car deck, which leads to the flight deck. A former 'heavy metal' pilot like myself feels immediately at home.

Roger is already running through checks from the right seat. Between him and the captain's seat is an airline-style throttle quadrant with four throttles and four propeller pitch levers. In front of the captain is a large fibre optic gyro (FOG) compass, airspeed indicator, turn and slip, and numerous gauges relating to prop pitch and pylon/rudder position. Both the captain and FE have a set of conventional dual controls, that is to say rudder pedals and

Cockpit of an SRN-4, which had three crew.

a control column. These, apparently, are made to designs originally intended for the SR-45 'Princess' flying boats of the 1950s. Behind the captain and FE is the panel of the FO, who looks rather like an air traffic controller sitting at a desk in front of two colour CRTs. Today he will be using the 3cm (1.2in) X-band radar. The 10cm (4in) S-band equipment is superior in conditions of heavy rain, but does not offer such good range and bearing penetration. With an anticipated cruise speed of 55 knots through the Strait of Dover – perhaps the busiest stretch of water in Europe – Peter will be talking to Linton regarding other traffic throughout the flight. The APUs (auxiliary power units) are already running, and as I settle in the jump seat and don my headset I am greeted by the familiar hum of the alternators. Linton and Roger discuss the deferred defects shown in the technical log. Nothing of significance is listed. Each flight crew member then runs through a long list of pre-flight checks, cross-checking important items between them. Linton calls for the main engine starting checks and requests start clearance from the hoverport's control tower. Dover Port Control is worked on the other box, as only the hoverport itself is within the tower's jurisdiction. 'Port Control, Hovercraft zero six for start proceeding the Western in three,' which elicits the immediate response: 'Hovercraft zero six start, report lifting.' Things now happen very quickly. Roger starts all four engines, starting with number 4. (The engines are numbered left to right 3,1,2,4, front pylons being 1 and 2, and the rear pylons 3 and 4. The engines power their respective pylons/lift fans.) Electrical power is channelled from the APU alternators, throttle levers are checked closed, and the friction nut set. No.4 engine is selected on the starter, and the engine stop button is reset. The number 4 low-pressure (LP) fuel valve is opened, and an inlet pressure of 20psi noted, before pressing the start button. Starting is then largely automatic, with the FE checking for a rise in compressor and turbine RPM, and that the jet pipe temperature (JPT) rises and stays within limits (630°C transient).

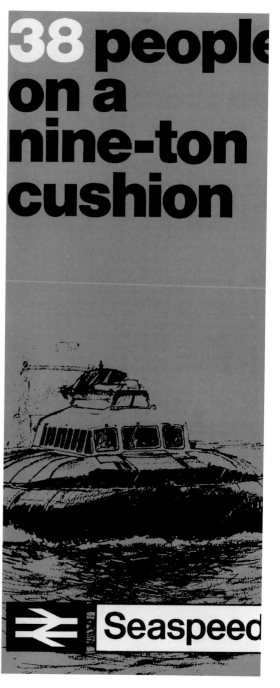

3. British Rail Seaspeed timetable.

Seaspeed

Hovercraft service

**Southampton
Cowes
and
Portsmouth
Ryde**

**with connecting trains
from and to London**

**13 September 197
to 2 May 1971**

hoverspeed

The Magnificent Flying Machine

Dover-Calais

When you take your car to France, get there in record time. The unique Hovercraft flies you from Dover to Calais in just 35 minutes, at speeds up to 65 mph. Skimming over the waves on a cushion of air, it's still the fastest and most exciting way to cross the Channel. On board, relax and enjoy personal service as our attentive cabin crew bring refreshments to your seat. With only 55 cars and 380 passengers, loading and unloading is fast and simple. Hoverspeed: for passengers who want to be treated as individuals.

For reservations and information, call 08705 240 241 or see your travel agent.

hovercraft seacat superseacat

4. Last Hoverspeed advert for the hovercraft.

Pre-operational Guide to the B.H.C. *SUPER-4* Hovercraft

5. BHC Super 4 brochure.

6. Pindair sales brochure.

Vol. 4 No 8 (New Series) November-December 1970   10s.   (50p.)

HOVERCRAFT WORLD

WORLD'S FIRST HOVERCRAFT JOURNAL

THE
1971 INTERNATIONAL
HOVERCRAFT
DIRECTORY

Incorporating the

1970 Annual Industry Review

7. Dedicated hovercraft magazine.

Hovercraft:
the Early Days

This leaflet sketches in the history of the Hovercraft between its invention by Sir Christopher Cockerell and the first cross-Channel passenger-carrying services, a crucial development period with which BP are proud to have been associated

8. BP oil brochure.

Air-Cushion Vehicles

November 1970   5s

The International Hovercraft Journal

Inside:

Rally Round-up
Commercial Operations 1970

Vol 4 No 7 (New Series) SEPTEMBER/OCTOBER 1970   3s. 0d

HOVERCRAFT WORLD

WORLD'S FIRST HOVERCRAFT JOURNAL

*Birds of a feather
at Calais Hoverport*

Featured Inside:

**HOVERMARINE
TRANSPORT PROGRESS**

10. *Hovercraft World* cover.

11. BP hovercraft booklet.

The Spark of Genius

A backward glance
in pictures of how
an idea caught
on

BP marine international

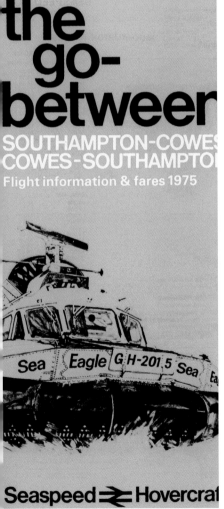

the
go-
between

SOUTHAMPTON-COWES
COWES-SOUTHAMPTON

Flight information & fares 1975

Sea Eagle GH-201.5 Sea Ea

Seaspeed ⇌ Hovercraft

12. Seaspeed Solent brochure.

Artists' visions of 1950s hovercraft.

Once the start is complete the compressor RPM is observed to stabilise at 5,500, the starter light checked out, and all temperatures and pressures – a plethora of which surround the FE ahead and to his right – are checked within limits. The procedure is then repeated for the other engines, ensuring that the tail doors are closed before starting engines 1 and 2 (the inboard engines).

The car deck supervisor will have produced a load sheet using standard vehicle/passenger weights. He calls the flight deck during the start with the figures, which are taken down by FO Peter Bird: 'We've got forty-two units, 252 passengers, we've got a dog on board, eighteen crew and one supernumerary.' A muffled 'thud' is felt as the nose ramp shuts. A 'telltale' flag appears over the nose. Another call comes through from the car deck a few moments later, 'Hello flight deck, ramp secure, thank you.' The captain calls for the pre-lift checks, which are read out by Roger, and responded to variously by all three flight crew. Linton calls the tower again, 'Six departure clearance please.' The lady controller answers, 'Six you have light air south easterly, you're all clear.' Then, following a quick flick of the radio selector, another radio exchange, 'Port Control zero six lifting.' 'Zero six proceed Western.' The FE opens the throttles from idle to smoothly achieve 10,500 compressor RPM, ensuring that the JPT remains within limits (510°C), and that the engines are not over-torqued. Linton handles the propeller pitches ensuring a static hover initially, and GH-2006 rises into the air. The sensation is not unlike hovering in a helicopter, although it is much smoother.

As we reverse off stand with GH-2007 to our left, I find it quite awe inspiring how manoeuvrable these giants are in skilled hands. By careful manipulation of the propeller pitches, rudder, and pylon angles, we move backward past GH-2007, then turn through

To double the SRN-4s capacity, it was given a 55ft mid-section stretch in 1977.

The giant rear bag of an SRN-4 mark 3.

180° before moving off toward the sea. Once forward motion is established, over-torquing is less likely and Linton calls, 'Cruise please.' Roger eases the throttles forward to achieve 10,100 turbine RPM, the primary power reference other than for close manoeuvring. In the meantime, Linton has started moving fuel around the craft using pumps and valves controlled from the overhead panel, in order to trim for optimum centre of gravity (C of G) in the prevailing conditions. During the initial turn on the apron, Peter has checked the 'green-white-green' lights at the harbour entrance through the flight deck's rear windows (that light combination means 'you may proceed with RIT clearance'), and announces, 'you have the signal.' We proceed past the pier that abuts the hoverport, past the main entrance to the harbour, and onto the busy Channel. Peter's voice comes over the intercom, 'Heading one zero five.' A third agency is then called, this time by Peter, 'Dover Coastguard, good afternoon, hovercraft zero six just departed the Western bound for Calais.' The Coastguard 'auto numbers' the craft on its radar and no more calls are necessary: 'Zero six good afternoon sir, good crossing.'

Throughout the flight, Linton continues to re-trim the fuel on the overhead panel, and Roger adjusts his throttles and the pitch levers to maintain cruise power. Linton has a windscreen wiper and fresh water washer button on the left horn of his column, and this is regularly used to good effect. Through the headset, Peter's calm voice warns of other traffic and appropriate headings to steer clear of it, 'Another southwester coming down to port, no problem seventy four just under four miles, tracking four and a quarter ahead of it. I'm going to keep you on one hundred for a while to keep you clear of a couple of Dover-bound ferries, the closest of the two being one twenty at seven.' Once Linton identifies a target through the flight deck windows (hence the importance of those washers) Peter

The Sea Cat was thought by many to be a replacement for the hovercraft – but it wasn't. Despite what many may think it wasn't the catamaran, nor the Channel Tunnel, that put an end to the cross-Channel hovercraft service, but the demise of duty-free shopping.

invites him to tighten up on it visually, and once it is passed, gives a new heading. We cruise at 55 knots (approximately 63mph) indicated, normal for a loaded Mark 3. During a quieter period, Linton explains how he flies the craft. Left or right pedal input turns the machine around the central axis by a combination of differential pylon and rudder movement, while turning the control column moves the craft bodily in that direction (i.e. the heading remains constant) by all four pylons turning in unison. When hovering, this is further augmented by differential propeller pitch. As we approach the wake of a Sea Cat, he demonstrates the integrated lift/propulsion system. By hauling back on the column, power is diverted to the lift fans at the expense of forward thrust, effectively lifting the craft over the waves. Nonetheless, a jolt is still felt on an otherwise glassy sea. Peter, who has little external vision from his panel, accurately comments on the intercom, 'I take it that was the Sea Cat's footprints!'

Ten minutes out and Peter calls the tower at Calais for the weather and landing information: 'Calais six, *bonjour*, with you in ten minutes for fuel please,' and is advised of the surface wind, visibility, and which stand to park on. 'The pad is bearing one two four at four and a half miles,' says Peter, and shortly after that we go visual, approaching on the extended centreline track of 170°. The hoverport is outside the harbour, and hovercraft do not communicate with Port Control. Roger runs through the arrival checks and as we approach the beach Linton asks for 'ten six front', whereby Roger sets 10,600 compressor RPM on engines 1 and 2 to slow the craft down before we leave the water. The flight manual stipulates a maximum transition speed from water to land of 35 knots (*c.*40mph), but across the sand extra care is taken to limit damage to the skirt. (At low water, half a mile of sand is crossed before arrival at the ramp.) Linton hovers the craft slowly up to the front of the terminal, which it towers over before the throttles are closed and we sink onto the seven landing feet with a slight bump. He calls the tower, 'Calais, six in position,' and Roger presses all four stop buttons to shut down the main engines. Both APUs continue running throughout turnarounds. A Jet-A1 bowser hooks up with the pressure refuelling point on the left of the craft. A Mark 3 holds 36,000 litres (13,260 gallons), and is normally refuelled to 24,000 litres at Calais for a round trip. It will burn about 5,850 litres (1,285 gallons) for the flight to Dover and back on a calm day.

I accompany Linton on his walk around inspection of the neoprene mesh and rubber skirt. It has a periphery of 117 rip-stopped fingers to contain damage. On average, twenty are replaced per week, at a cost of £300 each. Meanwhile, on the fuselage roof, FO Peter Bird is inspecting the massive propellers, and Linton explains that these are perhaps the Achilles heel of the hovercraft. Enough engines and spares are either owned or readily available to see the craft through another five years. However, the four-bladed props are unique. They are the largest airscrews in the world (22ft), and cost £96,000 *per blade* (four per head making sixteen per craft) to replace, with a lead time from BAE Systems of three years. To replace all would cost £1,536,000!

Inside, the economy-class cabins are much as one would expect to find on a regional airliner, and all seats are forward facing. The left-hand cabin includes the forty-seat business class section, more like the first-class carriage of a train, with club seating and plush tables. We regained the flight deck for the return trip. At Dover, the ramp to the apron is much steeper, and Linton explains that for 2 hours either side of low water a Mark 3 must fly its approach at at least 25 knots in order to reach the apron at heavy weights. In poor weather, the FO talks the captain between the harbour walls and onto the white line, a final approach track of 307°. In view of the confined nature of the flight path and required speed, a minimum of 300m (1,000ft) visibility for landings at Dover is applied. At Calais, zero-zero approaches are approved. I fly six sectors in all with Linton and his crew, each one routine yet with slight differences.

The flight times are 32, 31, 29, 28, 30 and 30 minutes respectively, and the last includes a night approach to Dover. Once all the passengers and vehicles are off, the flight crew (minus the FO) retake their seats to move the craft onto the maintenance jacks. Just two engines are started, quite adequate with the light craft. A neutral propeller pitch is set. Where cars had been marshalled earlier in the day, seven giant electro-hydraulic jack heads are visible in the tarmac. As the power is increased, the craft rises once more. Unseen, two huge fixed steel cables have been attached to the tail of the craft via a massive bridle. To the front, another two cables are connected to winches, which slowly start to ease GH-2006 toward the jacks. At flight deck level, a marshal atop a gantry speaks to the crew on the tower frequency. 'OK Linton, on the line, thirty, twenty, ten.' The information is a courtesy, as both Linton and Roger are passengers; the winch men are driving the craft and the marshal's wand signals are to them, not the flight deck. Then, a tug from behind as the bridled cables bite. The throttles are closed and the craft sinks on to its jacks, and the precise position is double checked against white witness marks on the skirt, then the engines are shut down. GH-2006 will be raised after our departure so that engineers can work overnight with the skirt clear of the tarmac.

Immediately beyond the marshalling gantry, cars are queued to board a Sea Cat on the adjacent link span. People rush forward with cameras, and the scene from the flight deck on what is my last ever SRN-4 flight is a sea of flashlights. Even after more than three decades, these massive hovercraft still turn heads.

Building the new hoverport.

A typical busy day in the life of the hover terminals.

Princess Anne's arrival at the hover terminal at Dover.

'Hovergirls' greet boarding passengers.

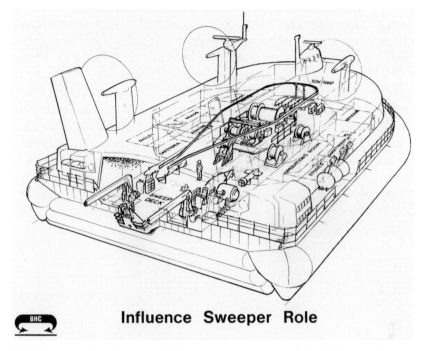

Influence Sweeper Role

This illustration shows how the SRN-4 was taken very seriously by BHC as a military craft from the outset due to its unique ability to glide over a mine with low-risk of detonation whilst carrying out mine-sweeping duties.

SRN-4 operated by Seaspeed skims the Channel at speed.

And so, at 20.35 on 1 October 2000, as Capt. John Hawkins shuts down the engines on *Princess Margaret* for the last time, *Princess Margaret* had completed 48,195 operating hours in her life. It is almost over. But not quite. Twenty minutes behind, GH-2007 *Princess Anne* is operating a special farewell trip for the loyal staff of Hoverspeed.

No longer constrained by economic operations, Capt. Nick Dunn opens up the throttles and achieves an airspeed of 75 knots (86mph!) over the choppy sea. Yes, this truly is the fastest way across the Channel! And as she approaches Dover, landing lights ablaze, Capt. Dunn slows to a crawl and gently inches GH-2007 forward through the dark and touches the beach where, in 1959, Sir Christopher Cockerell (who died in 1999) came ashore in the SRN-1, in a salute to her ancestors, and the man behind them. It is as if this gentle giant is saying 'thank you for the last thirty-two years.' Then she slowly withdraws to position on the hoverport's apron next to her sister. As she shuts down, an era draws to a close. Today, the English Channel grew wider again.

SEVEN

HOVERSPEED

Hoverspeed's projects captain, Captain Linton Heatley, was tasked with moving SRN-4 Mark 3s *Princess Margaret* and *Princess Anne* from Dover to lay up at HMS Daedalus. This is his account of the move:

After sourcing four different lay-up locations, with reasonable access and above average security as premium requirements, secure storage became available for the two hovercraft at HMS Daedalus, a former Royal Naval Air Station at Lee-on-Solent, Hampshire. HMS Daedalus is also the home of the Hovercraft Museum, and coincidentally the parking places of the two SRN-4 Mark 3s is such that, if no sale takes place, the *Princess Anne* will be gifted by Sea Containers to the nation through the Hovercraft Museum. However, the *Princess Margaret* may be scrapped.

The 108 nautical mile positioning journey from Dover to Lee-on-Solent was to be straightforward, on paper. After all, the craft had been built by the British Hovercraft Corporation at East Cowes, Isle of Wight, only 5 miles away across the Solent. For senior technician Mick Wells and me, however, the safe positioning of the world's two largest hovercraft to HMS Daedalus presented a unique challenge. There was the almost daunting steepness of the unfamiliar Lee-on-Solent slipway, with very little lateral room for the craft to cross a public road between street lamps and through solid steel security gateposts, and close turning between the hangars on the air station. Scale drawings and small manoeuvring tolerances said it could be done – the visual perception was a lot harder!

The complexity of the operation with such tight tolerances required selfimposed daylight-only limitations at HMS Daedalus. After legal delays relating to the ground lease had been overcome, significant and protracted weather periods over the southern half of England, the worst for many years, further delayed the deliveries. In consultation with Gosport Borough Council and the Hampshire Police, arrival at the base was planned to coincide with times of minimum disruption to the public, caused by the closure of the road running across the top of the slipway.

Hundreds of tonnes of beach shingle would require bulldozing, and later replacing, if the craft were to access the slipway at times other than high water, which would have been a very restrictive window with only limited winter daylight hours available.

The original plan called for slipway assistance using three large tow trucks from a local operator. However, Hoverspeed's Captain John Hawkins, who was in command for both journeys, modified the plan.

In carefully controlled but calculated manoeuvres, both hovercraft were taken directly up the ramp, the tow trucks only being used once the SRN-4s were safely inside the gates

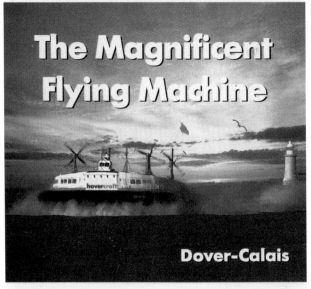

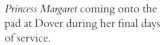

Princess Margaret coming onto the pad at Dover during her final days of service.

Advertisement for Hoverspeed.

Departing Dover for the 35-minute trip to France.

The newest livery aboard SRN-4 in 1999 with tables added.

– an unpractised and highly skilled manoeuvre. The first delivery, early in the morning of Saturday 3 December, left many long-standing Hoverspeed staff, who had gathered to watch *Princess Anne* depart Dover at 05.30, with nostalgic feelings of a funeral. This was to be a false start though, as 3 miles east of Folkestone the craft developed an oil cooler problem and returned. An omen? No, the successful delivery of the SRN-4 took place later that morning when technical repairs had been effected.

The final flight of *Princess Anne* as she enters her new home at the Hovercraft Museum.

Princess Anne sitting on the beach waiting to move into position to cross the main road and enter the museum site.

The sheer sight and aura of these giant hovercraft arriving and the uniqueness of the operation at Lee-on-Solent encouraged an army of onlookers and enthusiasts, many of whom had waited in vain, and frequently in rain, on previous days when local rumour had said that the SRN-4 were arriving.

One of the visits to islands off England and France – Goodwin Sands.

THE PRINCESS MARGARET

On Saturday 16 December, with great media interest and in an honouring farewell to many on the Isle of Wight, Captain Hawkins, with fuel, time and the weather on his side, deviated towards the end of the second delivery journey to take the *Princess Margaret* along the north-east coast of the island.

The craft staged a 'fly past' at Ryde, where Hovertravel's service to Southsea now represents the United Kingdom's last remaining commercial hovercraft operation. This was in order to pay tribute to many of the former British Hovercraft Corporation employees, before turning back to Lee-on-Solent and the completion of another safe, successful voyage. Neither craft was empty; between them they were carrying a lot of Hoverspeed's remaining stock of service spares on their vehicle decks. Yet the *Princess Margaret* averaged 48 knots on the trip around the south-east coast of England. Speed set these machines apart.

Heralded at the beginning alongside Concorde, but more recently operating in a highly competitive marine environment, so ended the 1960s brave quantum leap for the application of the hovercraft principle, leaving a wealth of safe technical and operational experience. Cross-Channel hovercraft attracted a cult following, akin to intercontinental flying boats and transatlantic airships. It seems unlikely that the world will ever again see a similar fully amphibious, high-speed form of marine transportation. Those that never saw the SRN-4 will never know what they missed. Was this the last time that one of these great giants of classic transport will fly?

EIGHT

MILITARY HOVERCRAFT

As we have seen, it has been largely the British that have dominated the world hovercraft scene but while the Isle of Wight was experiencing a new craze the rest of the world wasn't far behind it.

The USA had by now started to produce its own hovercraft, especially with the Marines taking a fancy to the idea of being able to run straight onto a beach at high speed. The Hydroskimmer was the USA's answer to an assault ship that could fly troops and equipment onto the beach. As with the British, the United States' Hydroskimmer was crafted by the hands of highly skilled aero engineers, Bell Aerospace, who had a record with hovering thanks to their legendary Bell 47 helicopter.

The Hydroskimmer weighed 35 tons, was nearly 65ft long and had a total power output of 4,000hp. The US Navy put their new mammoth hovercraft through a series of tests and found that it coped with everything it needed to do and more. It had an almost modular construction which enabled it to be converted or adapted into various configurations for stability depending on use.

It looked totally revolutionary compared to anything seen before with its large twin-ducted fans improving propulsion at the rear. One of the most obvious things about the Hydroskimmer was the lack of a skirt to support its air cushion. Despite this, the craft could hover half a metre from the surface and was the largest hovercraft built in the United States at this time.

At its first public appearance on 28 September 1964 the craft skimmed Lake Erie, one of the Great Lakes in North America, in front of many people who seemed to think it was from a future conflict and not of their time. It travelled at almost 80mph, loaded with thirteen fully equipped Marines ready to demonstrate their rapid deployment actions as the craft flew over land and water before hitting the drop zone.

The workings of the Hydroskimmer consisted of four inverted fans to provide lift, while their power source was internal. These series of initial tests proved just how valuable the vehicle and concept was to a fighting force such as the Marines. Its speed, manoeuvrability, payload and relatively low cost of operation ticked all the right boxes and all at a time when the United States needed to keep its crucial line of defence one step ahead of everyone else. That was the intention anyway, but as we will later see, the Soviets had a secret weapon up their sleeve.

In addition, it wasn't just Bell that were working on new air cushion projects. Republic Aviation, Grumman Aircraft, General Dynamics and Martin-Marietta, the very companies that excelled in producing multi-million-dollar supersonic aircraft and their operating systems, had by now also taken the hovercraft under their wing.

The US Hydroskimmer at speed.

One of the Hydroskimmer's lift fan units.

An artist's impression of the Hydroskimmer.

LCAC (LANDING CRAFT AIR CUSHIONED)

After the Second World War the US Army and Navy had to look at the future of launching an assault on beachheads. While the Normandy landings had demonstrated just how vital the landing craft is to an invasion force it was by now becoming dated, drawn back by its slow speed and vulnerability from air attack. It wasn't that the landing craft had become obsolete, far from it in fact as most navies across the world continue to use them; it was just the pace of warfare was threatening to become faster by the day which meant governments were constantly on the lookout for new vehicles, systems and technology that would ensure their nation could at least keep ahead.

The United States had proven their interest in hovercraft with the Hydroskimmer, which was only really meant to be a prototype. In the late 1960s a new concept was being pushed around the US Navy, a new type of hovercraft that while still being fully amphibious could also deliver a mechanised force to land far more quickly than any other craft had done so previously. The design was called LCAC – Landing Craft Air Cushioned. Bell Aerospace constructed a prototype which seemed to offer a very versatile platform. With an aluminium construction it almost looked like an extension to an air transport aircraft than anything marine based. What was visibly apparent was the large open cargo area which in previous military hovercraft designs had been covered which restricts on clearances. The LCAC offered a very generous payload capacity, between 60–75 tons, which meant that the hovercraft was now able to beach a main battle tank quickly and effectively.

A modern LCAC operated by the US Marines.

An LCAC hovers into the back of its assault ship.

A modular construction meant that areas of the craft could, if needed, be adapted to carry out a specific role.

During the advanced development stage, two prototypes where built. JEFF A was designed and built by Aerojet General in California. JEFF B was designed and built by Bell Aerospace in New Orleans, Louisiana. These two craft confirmed the technical feasibility and operational capability that ultimately led to the production of LCAC. JEFF B was selected as the design basis for today's LCAC.

The first LCAC was delivered to the Navy in 1984 and Initial Operational Capability (IOC) was achieved in 1986. Approval for full production was granted in 1987. After an initial fifteen-craft production competition contract was awarded to each of two companies, Textron Marine and Land Systems (TMLS) of New Orleans, and Avondale Gulfport Marine, TMLS was selected to build the remaining craft. A total of ninety-one LCAC have now been built. The final craft, LCAC 91, was delivered to the US Navy in 2001. This craft served as the basis for the Navy's LCAC Service Life Extension Program (SLEP). LCAC first deployed in 1987 aboard USS *Germantown*. LCAC are transported in and operate from all amphibious well deck ships. The craft operates with a crew of five. In addition to beach landing, LCAC provides personnel transport, evacuation support, lane breaching, mine countermeasure operations, and Marine and Special Warfare equipment delivery.

Today the LCAC is used to transport the weapons systems, equipment, cargo and personnel of the assault elements of the Marine Air-Ground Task Force from ship to shore and across the beach. LCAC can carry heavy payloads, such as an M-1 tank, at high speeds.

Roll out of the 160-ton craft at the NASA Michoud facility in Louisiana.

The hovercraft is essential to a fighting force, delivering cargo, fire power and supplies to a beach head over five times faster than a conventional ship.

The U.S. Navy skims the surface with the LCAC from Textron Marine Systems – the leaders in air cushion technology.

TEXTRON

The American LCAC programme has been developed by Bell Aerospace Textron from the start.

Firepower being hovered to a beach head.

The LCAC payload capability and speed combine to significantly increase the ability of the Marine Ground Element to reach the shore. Air cushion technology allows this vehicle to reach more than 70 per cent of the world's coastline, while only about 15 per cent of that coastline is accessible by conventional landing craft.

LCACs have been used in peacetime missions, delivering aid to victims of the Haiti earthquake in 2010 and rescuing people during the New Orleans floods. The hovercraft continues to display its abilities that no other craft can match around the world in saving, helping and rescuing civilians. As is commonly said by the US LCAC crews: 'No beach is out of reach.'

LCAC

Builder: Textron Marine and Land Systems/Avondale Gulfport Marine
Date Deployed:1982
Propulsion: Legacy: 4 Allied-Signal TF-40B gas turbines (2 propulsion/2 lift); 16,000hp sustained; 2 shrouded reversible pitch airscrews; 4 double-entry fans, centrifugal or mixed flow (lift)
SLEP: 4 Vericor Power Systems ETF-40B gas turbines with Full Authority Digital Engine Control
Length: 87ft 11in (26.4m)
Beam: 47ft (14.3m)
Displacement: 87.2 tons (88.60 metric tonnes) light; 170–182 tons (172.73–184.92 metric tonnes) full load
Speed: 40+ knots (46+mph; 74.08km/h) with full load
Range: 200 miles at 40 knots with payload/300 miles at 35 knots with payload
Crew: Five
Load: 60 tons/75 ton overload (54.43/68.04 metric tonnes)
Armament: 2 12.7mm MGs. Gun mounts will support: M-2HB .50 cal machine gun; Mk-19 Mod3 40mm grenade launcher; M-60 machine gun.
Electronics: Radars, Navigation: Marconi LN 66; I band/Sperry Marine Bridge Master E.

ZUBR CLASS

The vast Zubr class hovercraft is of Soviet design. It weighs in at 550 tons and is currently the heaviest hovercraft in the world. There are currently nine craft built, just a few of which are in service around the world. The Zubr is used by the Russian, Ukrainian and Greek navies. The transfer of the *Kefalonia* (L-180), the first of two Zubr hovercraft purchased by the Greek Navy, marked the first time that a Russian-made craft was used by the navy of a NATO member.

It is designed to meet a requirement, much the same as the US LCAC craft, by landing assault units (such as Marines or tanks) from equipped/non-equipped vessels to non-equipped shore, as well as transport and plant mines.

High strength and floatability of the craft are provided by a rectangular pontoon, the main load-carrying part of the ship's hull. The superstructure built on the pontoon is divided into three compartments with two longitudinal bulkheads, combat material compartment in the midsection fitted with tank ramps, and outboard sections housing main and auxiliary propulsion units, troop compartments, living quarters, and NBC protection systems. To improve working conditions in the battle stations, troop compartments and living quarters are fitted with air conditioning and heating systems, sound/heat insulating coatings, and structures made of vibrodamping materials. The craft provides suitable conditions for the crew to take meals and rest.

The Russian Zubr class is the world's largest military hovercraft.

A Zubr unloading vehicles, firepower, infantry and cargo.

Personnel are protected against effects of weapons of mass destruction with airtight seal-ing of combat stations, crew and troop compartments, as well as individual gas masks and protection suits. The ship is also protected from influence mines with the horizontal wind-ing to compensate for the ship's and the transported material's magnetic fields. The central command post and MS-227 device compartments are strengthened with alloy armour.

The Zubr has a cargo area of 4,300 sq. ft, and a fuel capacity of 56 tons. It can carry three main battle tanks up to 131 tonnes, or ten armoured personnel carriers with 140 troops up to 115 tonnes, or up to 500 troops with 360 troops in the cargo compartment. At full displacement the craft is capable of negotiating up to 5° gradients on non-equipped shores

The Zubr is one of the world's most armoured hovercraft. It is a rare sight to see them in action but they are ready and waiting when needed.

and 1.6m-high vertical walls. The Zubr remains seaworthy in conditions up to sea state 4, while the craft has a cruising speed of 30–40 knots.

Operators:

Russian Navy:	770 *Evgeny Kocheshkov* (former MDK-118); 782 *Mordovia* (former MDK-94)
Ukrainian Navy:	*Donetsk* (former U 420); *Artemivsk* (former MDK-93, U 424)
Hellenic Navy:	Operates four craft: three were commissioned in 2001 – *Kefalonia* (L180) was donated from the Russian Navy and upgraded; *Ithaki* (L181) which was completed in Ukraine; and *Zakynthos* (L183) which was built in Russia. A fourth craft, *Kerkyra* (L182), was launched in June 2004 at St Petersburg yard and was commissioned in January 2005.

VOSPER THORNYCROFT

I wonder if John Thornycroft could ever had imagined that one century after his original Victorian experiments his very company would go on to build large hovercraft for the military? The fact is true and was the result of a joint venture when in 1966 Vosper Ltd and John Thornycroft & Co. Ltd merged to form Vosper Thornycroft Ltd. Like Thornycroft, Vosper were also boat builders, in particular large high-powered air-sea rescue launches and general pinnaces for the RAF and Admiralty. A number of these boats by both firms can be seen today in various states. Two preserved examples lie at the RAF Museum in Hendon while others have been converted into house boats such as those at Bembridge and also in Shoreham-by-Sea, West Sussex, where they were broken up after the war. The

VT-1 on trial with Hovertravel but never put into service.

The VT-1 shows of its semi-amphibious role as it unloads its cargo and passengers at the Southsea Hovertravel terminal.

traditional method of mahogany cross planking was the most common form of hull construction. Vosper Thornycroft produced and developed several hovercraft but the most notable of their efforts were the VT-1 and VT-2 which began trials in 1969.

Weighing in at 77 tons, VT-1 was powered by two Lycoming TF-20 Marine Gas Turbine engines and had a cruising speed of 40 knots. It was the second commercial hovercraft that could carry people and vehicles, 148 seated passengers and ten average-sized cars.

VT-1 was 95ft long with a beam of 45ft. It was quite strange: while it may have looked like a hovercraft and had a clever clean peripheral skirt, it lacked all the fully amphibious qualities that make a hovercraft appealing. It also had two underwater rudders which acted in the same way as on a conventional boat. This was because it lacked the form of external propulsion and instead used a water screw to propel it along. On the face of it, it may have looked like VT-1 designers had missed the point of the whole hovercraft concept, but when you look closer, it had a definite advantage for certain roles. Its two propellers initially made it more manoeuvrable at low speeds and when entering a busy harbour in strong winds this is very favourable for a hovercraft. For a start the propellers were a lot quieter compared to the noisy fans of gas turbine engines and their propellers. This meant relatively low effort was required for propulsion and smaller engines than normal were selected given the size of the craft. VT-1, unlike fully amphibious hovercraft, could corner without struggling to avoid skidding sideways. The Lycoming engines had not been used in hovercraft before and the result was that VT-1 was a lot more economical to run with its conventional type of marine engines.

The craft performed well up to expectations, and turned out to have remarkably good sea-keeping characteristics. In conjunction with the Department of Trade and Industry (DTI), who were encouraging the hovercraft industry with sundry financial support, Vosper Thornycroft carried out a series of trials during the winter of 1970/71. Based in the Channel Islands, these encountered a good proportion of rough weather, and it was shown that the VT-1 could maintain a speed of 26 knots in 12ft waves.

The VT-1 was designed specifically for commercial use, but the company also had an eye on military applications. The steadiness of the VT-1 at speed in a rough sea, compared with a conventional boat, had obvious attractions when applied to a weapon platform. This could be convincingly, if frivolously, demonstrated to potential customers by inviting them to drink a glass of gin and tonic whilst proceeding at speed in conditions which, in a conventional patrol boat, would have required them to use both hands to hang on for dear life!

There also seemed to be plenty of potential for civil hovercraft. Technically, the performance of the VT-1 prototype was promising, and two production craft were laid down on the strength of active interest displayed by Hovertravel Ltd, who were building up a lot of operating experience with the smaller SRN-5s and SRN-6s built by the British Hovercraft Corporation.

Unfortunately, despite various schemes, Hovertravel never acquired a VT hovercraft. A draft agreement to charter VT-1 number 001 (the prototype) for operation in the Channel Islands came to nothing, and efforts were redoubled to sell the craft elsewhere. In April 1971 the board were discussing the possibility of selling the two production craft (002 and 003) in the USA, in conjunction with a licensing agreement for building more craft in the States, but this too failed to materialise, and by May a hard look was being taken at halting further expenditure on development work; for example, on running the VT-1(M), a man-carrying model which was used for various experiments including, latterly, trials with water jet propulsion.

The interior of VT-1.

Cockpit of the VT-1 hovercraft.

Towards the autumn, proposals for a joint operation involving Rederi Aktiebolaget Centrumlinjen of Sweden were beginning to crystallise at last. VT was urgently in need of some solid operational experience with the craft, and there was a clear intention to subsidise the operation with a view to achieving this as soon as possible.

At the same time, the company had a firm policy to 'go equally hard for both commercial and military applications of hovercraft', and there was a proposal to allow the Navy Hovercraft Trials Unit (NHTU) to evaluate VT-1.

In October John Rix went to Malmo to sign the promotion agreement with Rederi Centrumlinjen and thereafter things moved ahead. Vosper Ltd, as VT's parent company, agreed to the subsidising of the operation, and a joint operating company, Centrumsvavarna AB, was set up to operate VT-1 002 and 003 between Malmo and Copenhagen, fitted out to carry passengers only. The craft themselves were sold to a finance company and leased back for this purpose. The service began in March 1972, and technically was a great success, the craft performing well and reliably. The first week or so of operation showed an encouraging load factor of 50–60 per cent, but then the state-owned opposition, operating in direct competition with hydrofoils and conventional ferries, cut its fares. The battle was on. A glance at the statistics tells the story only too plainly: the regular commuters could not afford to refuse the cheaper fares, and the average load dropped quickly to around 30 per cent of capacity during the week, but regularly shot back up again at weekends when Centrumsvavarna seemed to get a better share of the market. Weekend trippers were presumably prepared to pay more than regular travellers for a faster, more spacious and comfortable ride. But despite running two craft and twenty-eight scheduled trips per day, the statistics improved only slightly. With the end of summer in sight, and a probable reduction in the weekend demand, Hovertravel were asked to survey profitability and advise. By October the operation was at an end and the two craft returned to Portchester, having ferried nearly 200,000 people over the route.

A loss of about £500,000 resulted from the ferry operations, but a great deal of good experience and confidence was gained. The board was, not surprisingly, unwilling to repeat a similar operation which could have provided very little more in the way of worthwhile experience, and one or two such projects were abandoned in the early stages of discussion. It was agreed that further ferry operations could only be contemplated if the craft were actually sold.

Negotiations in April 1973 with British Rail, aimed at operating the craft on the latter's Isle of Wight ferry route, were not promising; calculations suggested that even at the sale price proposed (£600,000 per craft; not a particularly high figure) an operating loss would result. The board considered lowering the price still further, but in the end decided against it.

The salesmen continued to try very hard to find homes for the two VT-1 passenger craft, and in the autumn of 1973 there were still hopes of doing so, either on a route between Southampton and Cherbourg; from the mainland of Italy to Sardinia; or in Hong Kong. Alas, none of these ever materialised and eventually the craft were scrapped. People driving down the newly opened M27 motorway into Portsmouth were for some months able to view the sad remains in Harry Pound's scrapyard, where they were painted grey to conceal them as much as possible!

In 1972 hopes for hovercraft were still high. Considerable importance was attached to the future of military hovercraft designs since the Vosper Thornycroft board recognised that, whereas they had once enjoyed a significant technical lead, which almost amounted to a monopoly, in the design of the fast gas-turbine-powered torpedo boats, there was much more competition in the patrol boat markets now being addressed. Investing in hovercraft development was seen as one way of re-establishing a niche. Although hovercraft eventually turned out to be a dead end, this could not possibly have been foreseen at the time.

The Thornycroft works at Portchester. Aside from the two hovercraft, there are also various RAF and naval air-sea rescue launches and MTBs moored close by, a type of craft that Thornycroft had excelled in building.

A small Austin A40 boards the VT-1 for a cross-Solent adventure.

A demonstration of the VT-1's evacuation procedure.

Just before Christmas 1972, therefore, the Vosper board agreed on expenditure to convert the prototype VT-1 001 to a fully amphibious craft, which was to be known as VT-2. The new craft was a VT-1 with the skegs and water screws removed, and above-water propulsion substituted. The latter was novel, a good deal of thought having gone into its design.

Firstly, two ducted fans, rather than conventional airscrews were used. They had a relatively low rotational speed and were mounted above the stern, enclosed in annular ducts, which also housed pitch and yaw control surfaces. Drive shafts for these fans emerged direct from the gearboxes in the hull, each of which also drove lift fans; no additional gearing was included to make the thrust line horizontal.

The fans therefore were mounted at a slight angle to the horizontal, but the loss due to this was more than compensated for by the lack of an additional expensive and power-consuming gear train, to turn the fan thrust line horizontal. The fans themselves were specially designed and built by Dowty Rotol, and had variable pitch control. This meant that the thrust could be varied without changing the speed of rotation, or, therefore, the speed of the lift fans. Also, at low forward speeds, the craft could be steered by differential thrust alone. This made it, for an air-propelled craft, very quiet and very manoeuvrable. To drive each of the two propulsion fan/lift fan units, a Proteus gas turbine was fitted in place of the smaller Lycoming. The resulting craft was fully amphibious, with a slightly better payload capability than VT-1, and capable of 60 knots, which speed it could maintain comfortably in a sea state which would have been most uncomfortable, to say the least, in a Brave class FPB at full speed.

The hovercraft also compared well with conventional boats in its ability to withstand damage; this fact was perhaps not widely appreciated, the uninitiated tending to regard the flexible skirt as akin to a tyre, or even a balloon, to which one bullet hole might prove disastrous. In fact, quite large portions of the skirt and its peripheral fingers could be shot away without having any serious effect upon performance, and damage (arising in practice from wear and tear rather than bullet holes!) was quick and easy to repair, without any special facilities.

The VT-2 was another technical success, in that it performed well up to expectations. It was at that time the largest naval hovercraft in the world, with a payload of up to 33 tonnes. It proved popular with the Navy Hovercraft Trials Unit (NHTU), who first hired the prototype for evaluation. Some exciting trials were carried out with the Army, on exercises in the Hebrides, when troops and vehicles were delivered rapidly to some most unexpected places. With this encouragement, Vosper Thornycroft designed various different configurations for the export market. The Logistic Support Hovercraft was the basic prototype configuration. In this role, it could carry 130 fully equipped troops, together with transport. The latter could comprise four Land Rovers with trailers, or a couple of 4-tonne Bedford trucks, or even three Scorpion light tanks. The Fast Missile Hovercraft had exactly the same raft and propulsion system, but fitted with a 57mm or 76mm gun and two surface-to-surface missiles.

There was also a great deal of interest in the application of hovercraft to mine counter-measures, largely because, sitting on a big bubble of air, the craft proved almost invulnerable to underwater explosions unless detonated immediately beneath it. A considerable amount of study work was performed on this role, and after a refit, the MoD eventually bought the VT-2 prototype for extended evaluation as a support vehicle.

However, the world remained suspicious of hovercraft, unnerved, perhaps, by the unusual operating characteristics. Most small navies (and indeed, some large ones) tend to be very conservative and wary of adopting new ideas unless given a strong lead by major

The VT-2 operated by the Royal Navy featured ducted fans which meant a reduction in noise.

armed forces elsewhere. There are few rewards for initiative, especially in peacetime, promotion being dependent mainly upon time and a 'clean' record. Association with an unsuccessful project or incident are, however, quite enough to eliminate, permanently, all chances of promotion for the individual deemed responsible.

Despite the spectacular top speed of the hovercraft, some were put off by the low endurance (typically about 300 miles), compared with a conventional patrol boat. It was not always clear why this endurance was necessary: when patrolling a stretch of coastline a boat has little choice but to keep moving up and down over the same area at a slow cruising speed; stopping usually results in most unpleasant rolling. For example, on an anti-smuggling patrol lasting several days, sufficient fuel is required to keep underway, and this translates into quite a substantial number of nautical miles. A hovercraft is quite unable to carry enough fuel to cover the same distance at a slow cruise, and yet might be at least as effective since it can stop and drift for a while, without suffering such uncomfortable motion, thanks to its broad beam. It can also go ashore easily on almost any beach, and thus be relatively well concealed, more comfortable for the crew, but ready to move at high speed at short notice.

There were other roles, including mine hunting, where the unique capabilities of the hovercraft appeared to offer advantages to those with sufficient imagination, but even today there has been only limited acceptance of such craft. Despite the success of the IHTU trials, the British Armed Forces never showed much interest in hovercraft, and IHTU was eventually axed in a round of budget cuts. VT-2 finally followed her sisters to the scrap yard.

Without a lead by the Royal Navy or the British Army, the chances of overseas sales were further diminished, and despite some determined efforts at marketing both the VT-2 and other hovercraft designs abroad, the hovercraft department was eventually run down and most of the staff absorbed into the shipbuilding departments. VT-1 and VT-2 would have been splendid solutions, if the right problems could have been found to solve.

BH-7 WELLINGTON CLASS

Wellington class BH-7 was designed and built for the purpose of military and naval roles right from the very beginning. The first example achieved its maiden flight in the Solent in November 1969, only ten years after the original SRN-1 had first demonstrated to the world a hovercraft. Up until this point the BH-7 was the largest hovercraft operated by the British Forces. BH-7 was a 55-ton craft with a length of 78ft, while the skirt alone weighed 7 tons! This new craft could operate in 4ft waves at speeds of up to and sometimes over 70 knots. Fuel consumption was 1 ton per hour which gave the craft an approximate sortie time of 8 hours.

Power came from the tried and tested Rolls-Royce Marine Proteus (4250hp) which was beginning to become quite a popular engine for hovercraft of the '60s and '70s era. Lift was generated via a 12ft centrifugal lift aluminium fan, while propulsion was via a single large variable pitch propeller mounted on a pillion. Additional power was provided by two Rover gas turbine engines which were used to power the crafts electrical and operating systems. On its cushion, the height from the ground to the very tip of its propeller is 37ft, and some features of its design were similar to the SRN-4, such as the roof-mounted

The Wellington class BH-7.

An artist's impression of the BH-7.

cockpit which provides generous visibility for her crew. Other parts were directly shared with the SRN-4, including the pillion, windows and air intake assembly. Construction of the hull was made from a costly corrosion resistant light alloy, while the bow structure used a plasticell construction that was covered in strong high grade fibreglass. The hull consisted of a rectangular buoyancy tank divided up into watertight cells to provide floatation when the craft was resting on the surface of the water. It also formed a rigid platform around which the rest of the craft was built.

Like other previous military craft, BH-7 was designed to carry vehicles and troops. A Bedford 4-ton Army lorry would just about fit inside along with seating for soldiers and their equipment, or two Land Rovers could be accommodated.

In 1982–83 the Royal Navy craft was equipped with mine detection systems and sonar. The preserved example at the Hovercraft Museum still has its sonar tube in place. During its time under mine detection trials the craft went through a range of trials, operating out of the Royal Navy Air Station at Portland.

The second and third craft, designated Mark 4, and a further four examples of the Mark 5A, saw service with the Iranian Navy and were used publicly in 2011.

As mentioned before, it was the formation of the Interservice Hovercraft Trials Unit that would explore every possible scenario in which the air cushion vehicles could be put to use within MoD operations. Without the IHTU then the hovercraft would have continued to linger around while officials within the services continued to argue among one another over what it really was.

However, like the hovercraft itself the IHTU's very existence was owed to only a few forward-thinking men. As we have learned, it was actions of Lord Mountbatten that greatly assisted Sir Christopher Cockerell in getting the hovercraft from model to working concept with the SRN-1. It may come of no surprise to learn then that it was Lord Mountbatten that insisted a trials unit should be established from Gosport from where the hovercraft would be evaluated further over the subsequent twenty-one years.

A BH-7 craft with its sonar tube.

A BH-7 craft returns to the factory at Cowes.

The BH-7 craft proved durable and showcased an idea that the use of a high-speed patrol craft could have a vital role in homeland defence.

On 21 September 1959 Lord Mountbatten attended a display of the SRN-1 from HMS Daedalus with Vice Admiral Sir Walter Couchman and it was from here that they came up with the idea of the IHTU. The unit was firmly established at Lee-on-Solent in September 1961 under the command of Lt Cdr F.A.R. Ashmead, Royal Navy from HMS Daedalus, although at that time it was called HMS Ariel.

During the first three years of actions, the IHTU sourced craft by hiring them from their manufacturers for various tests and evaluations. From January to June 1963 the IHTU hired SRN-1 from Saunders Roe in order for the unit to conduct various tests and form training ideas for new pilots. By this time in the experimental craft's life, it had already been fitted with an 18in skirt, aiding its hover height.

The craft was fitted with sophisticated technology that would record data from its operations at sea. The IHTU worked closely with what was then named the Admiralty Experimental Works (AEW). During the craft's time with the unit at Lee-on-Solent, it went through various trials and tests at another site a stone's throw away from the base. The close proximity of the Army Camp at Browndown, with its direct beach front, made it the perfect location in which to hold a demonstration of its capabilities. Shingle was bulldozed to form jumps, obstructions and turns over a course that the craft would have to complete. Eventually, and with a further 52 hours on the clock, SRN-1 was returned back to Saunders Roe on 15 June 1962.

Things changed for the IHTU when in 1964 they took delivery of their first purchased craft made especially for the purpose of military intention. This was to be the first military hovercraft order. The craft was SRN-3, which shared much of its design with the commercial passenger-carrying SRN-2.

Artwork depicting the future of the craft.

BHC had visions of adapting the craft for a fast attack role and this illustration demonstrates how it could have looked.

The classic hovercraft shot, a BH-7 passes the Needles on the Isle of Wight.

NINE

WING IN GROUND EFFECT: THE AGE OF THE EKRANOPLAN

While all this was going on there was another force that was taking the hovercraft principle a step further and it certainly wasn't a friendly one. The mighty Soviet Union was, like the US and Great Britain, in a state of nuclear readiness in the late '60s. On an almost daily basis the media would report on Soviet activities, whether it be demonstrations of their capabilities or tales of things to scare us all in the West. Away from the eyes and radar of the US and its allies, the Russians were secretly developing a new vehicle, not from the confines of a test track but the vastness of the Caspian Sea.

However, what they were up to didn't escape the spy satellites of the United States when a picture was snapped of a naval base within the Caspian Sea. When American intelligence studied the picture they noticed a strange aircraft that looked unfinished. After much head scratching they worked out that whatever it was wouldn't fly even if it was a finished aircraft. Without looking for it they had found one of the Soviets' most top-secret projects of all time. They knew that they were looking at something very new and very hush hush.

It was called an Ekranoplan, a Russian word meaning 'skimmer'. The Ekranoplan is still a thing of wonder; it looks almost frightening now, so imagine how it must have been seen by the MoD and the Pentagon at the height of the Cold War. Like the hovercraft, it seems to be some kind of a cross between an aircraft and a boat, yet this one is definitely more aircraft. However, unlike the hovercraft it doesn't need a skirt to function, instead using small wings which are angled towards the surface with an anhedral wing configuration. The Ekranoplan works by starting out in the water, much like a flying boat. As the speed increases the unique aerodynamic shape of the wing creates a cushion of high-pressure air between itself and the surface, creating lift – this is called 'ground effect'. The Ekranoplans are commonly referred to as WIGs (Wing In Ground Effect) or GEVs (Ground Effect Vehicle). A huge advantage over the hovercraft is that they achieve far greater speeds and are purely propelled by aviation jet engines. However, WIGs are not amphibious like the hovercraft; they can't land on anything but water and when they do land it's at aircraft-like speeds.

Also known as a flarecraft, sea skimmer, Skim Machine, or wing-in-surface-effect ship (WISE), a ground effect vehicle (GEV) is sometimes characterised as a transition between a hovercraft and an aircraft, although this is not technically correct. Whereas a hovercraft is supported upon a cushion of pressurised air (from an on-board downward-directed fan), the principal effect of the proximity of the ground to a lifting wing is not to increase

A home-built WIG
demonstrates its aircraft
chartists.

its lift but to reduce its lift-dependent drag. Some GEV designs, such as the Russian Lun and Dingo, have used 'power assisted ram'-forced blowing under the wing by auxiliary engines to achieve a hovercraft-like effect or to assist the take-off. A GEV differs from a conventional aircraft in that it cannot operate efficiently without ground effect, and so its operating height is limited relative to its wingspan. Some GEVs are, in fact, able to climb out of ground effect.

The background on the wing in ground effect theory can be traced back before the Second World War. It was no accident that by the 1920s the ground effect phenomenon was well known, as pilots found that their aeroplanes appeared to become more efficient as they neared the runway during a landing operation. In 1934 the US National Advisory Committee for Aeronautics issued Technical Memorandum 771, Ground Effect on the Takeoff and Landing of Airplanes, which was a translation into English from French of a summary of research up to that point on the subject. The French author Maurice Le Sueur had added a suggestion based on this phenomenon:

Here the imagination of inventors is offered a vast field. The ground interference reduces the power required for level flight in large proportions, so here is a means of rapid and at the same time economic locomotion: Design an airplane which is always within the ground-interference zone. At first glance this apparatus is dangerous because the ground is uneven and the altitude called skimming permits no freedom of manoeuvre. But on large-sized aircraft, over water, the question may be attempted.

Small numbers of experimental vehicles were built in Scandinavia, particularly Finland, just before the Second World War. By the 1960s, the technology started to improve, in large part due to the independent contributions of Rostislav Alexeyev in the Soviet Union, and the German Alexander Lippisch, working in the United States. Alexeyev worked originally as a ship designer whereas Lippisch worked as an aeronautical engineer. The influence of Alexeyev and Lippisch is still noticeable in most GEV vehicles seen today.

The programme, led by Alexeyev, was organised by the Soviet Central Hydrofoil Design Bureau (CHDB), the centre of ground-effect craft development in the USSR. The military potential for such a craft was soon recognised and Alexeyev received support and financial resources from Soviet leader Nikita Khrushchev. Some manned and unmanned prototypes were built, ranging up to 8 tons in displacement. This led to the development of the 'Caspian Sea Monster', a 550-ton military Ekranoplan of 240ft (73m) length. Although it was designed to travel a maximum of 3m (9.8ft) above the sea, it was found to be most efficient at 20m (66ft), reaching a top speed of 350mph, as well as exceeding 460mph in test flight. The Soviet Ekranoplan programme continued with the support of Minister of Defence Dmitriy Ustinov. It produced the most successful Ekranoplan so far, the 125-ton A-90 Orlyonok. These craft were originally developed as high-speed military transports, and were usually based on the shores of the Caspian Sea and Black Sea. The Soviet Navy ordered 120 Orlyonok-class Ekranoplans. But this figure was later reduced to fewer than thirty vehicles, with planned deployment mainly in the Black Sea and Baltic Sea fleets. A few Orlyonoks served with the Soviet Navy from 1979 to 1992. In 1987 the 400-ton Lun-class Ekranoplan was built as a missile launcher. A second Lun, renamed Spasatel, was laid down as a rescue vessel, but was never finished.

Minister Ustinov died in 1985, and the new Minister of Defence, Marshal Sokolov, effectively stopped the funding for the programme. Only three operational Orlyonok-class Ekranoplans (with revised hull design) and one Lun-class Ekranoplan remained at a naval base near Kaspiysk. The two major problems that the Soviet Ekranoplans faced were poor longitudinal stability and a need for reliable navigation. Since the fall of the Soviet Union, Ekranoplans have been produced by the Volga Shipyard in Nizhniy Novgorod.

The Soviet aim with their Ekranoplans was to be able to deliver a fighting force with troops, vehicles or even launch a war head at the speed of a long-range jet bomber but all under radar. This presented a significant threat. However, at a time where everyone was watching each other there would have been a warning build up to this occurring, which thankfully never happened.

One particularly large example of an Ekranoplan was nicknamed the Caspian Sea Monster: at 550 tons it was larger than a B-52 bomber and weighed more than a Boeing 747. This giant could achieve 350mph at 60ft. At first sight it seems almost as if the designers just stuck the engines all at the front until they thought about the rest.

An Ekranoplan at speed in shallow water in a state of ground effect.

These craft would have operated under a specially formed squadron. To get some idea of the scale there would be anything up to 100 of these craft flying across an ocean carrying everything needed to invade a country without radar detection. By the time that country would be aware of the advancing Soviet Ekranoplans it would be too late. The last Ekranoplan to be built was the 400-ton Lun in 1987. It could carry six large long-range missiles and main battle tanks in its hold.

It wasn't until the collapse of the Berlin Wall and the fall of the Communist regime that the mighty Soviet military machine would end. The Ekranoplans were abandoned as funding ran out and the world started to hear of these once top secret machines in the early '90s. Even today there are still sensitive details we don't know. At the time of going to press the Russian Government has renewed the Ekranoplan project and is currently designing and developing new craft.

I personally feel that the Ekranoplan was one of the deadliest weapons never to have been used but like with most wartime technical advances its technology has many uses in civilian and commercial life.

Most GEVs developed since the 1980s have been primarily smaller craft designed for the recreational and civilian ferry markets. Germany, Russia and the United States have provided most of the momentum with some development in Australia, China, Japan and Taiwan. In these countries, small craft with up to ten seats have been designed and built. Other larger designs as ferries and heavy transporters have been proposed, but have not been carried to fruition.

After the collapse of the Soviet Union, smaller Ekranoplans for non-military use have been under development. The CHDB had already developed the eight-seat Volga-2 in 1985, and Technologies and Transport developed a smaller version by the name of Amphistar.

In Germany, Lippisch was asked to build a very fast boat for American businessman Arthur A. Collins. Lippisch developed the X-112, a revolutionary design with reversed delta wing and T-tail. This design proved to be stable and efficient in ground effect and even though it was successfully tested, Collins decided to stop the project and sold the patents to a German company called Rhein Flugzeugbau (RFB) which further developed the model.

Hanno Fischer took over the works from RFB and created his own company, Fischer Flugmechanik, which eventually completed two models. The Airfisch 3 carried two persons, and the FS-8 carried six persons. The FS-8 was to be developed by Fischer Flugmechanik for a Singapore-Australian joint venture called Flightship. The company no longer exists but the prototype craft was bought by Wigetworks, a company based in Singapore, and renamed as AirFish 8. In 2010 that vehicle was registered as a ship in the Singapore Registry of Ships. It is the first WIG craft to be flagged with the SRS, which is one of the world's ten largest ship registries.

German engineer Günther Jörg, who had worked on Alexeyev's first designs and was familiar with the challenges of GEV design, developed a GEV with two wings in a tandem arrangement, the Jörg-II. It was the third, manned, tandem airfoil boat, named 'Skimmerfoil', which was developed during his consultancy period in South Africa. It was a simple and low-cost design, but has not been produced beyond a prototype. The consultancy of Günther Jörg was founded with a fundamental knowledge of wing in ground effect physics, as well as results of fundamental tests under different conditions and designs that began in 1960. In 1984 Jörg received the Philip Morris Award. In 1987, the Botec Company was founded.

Today the GEV is being taken quite seriously by numerous engineering companies across the world, many with ideas of developing a super ship which will carry vast amounts of cargo around the world's oceans at a fraction of the cost of conventional air cargo but at aircraft speeds and shipping volume.

In recent years there has been much interest in prototype advanced military WIG projects for the US Marine Corps. *Manta* was a proposal that would see a fighting force reach a beach head far quicker than anything else in the past has ever done while being efficient and versatile enough to meet all the other strict demands of the Marines. Its stealth-like appearance makes us wonder if this will become the shape of things to come as the concept claimed a 60-ton payload in a sea state 5 and while cruising at 350mph.

Besides the development of appropriate design and structural configuration, special automatic control systems and navigation systems are also being developed. These include special altimeters with high accuracy for small altitude measurements and also lesser dependence on weather conditions. After extensive research and experimentation, it has been shown that 'phase radio-altimeters' are most suitable for such applications as compared to laser, isotropic or ultrasonic altimeters. However, only time will tell if this technology works in prototypes, which may lead to commercial and passenger operations.

Bob Windt of Universal Hovercraft in America developed the first flying hovercraft, a prototype of which first took flight in 1996 on the Mississippi River, near Cordova, Illinois. Since 1999 the company has offered plans, parts, kits and manufactured GEV hovercraft called the Hoverwing.

In Singapore, Wigetworks has also partnered with National University of Singapore's Engineering Department to develop higher-capacity WIG craft. Iran deployed three squadrons of Bavar-2 two-seat GEVs in September 2010. This GEV carries one machine gun and surveillance gear, and reportedly incorporates stealth technology.

As with the hovercraft in its early pioneering days one difficulty which has delayed GEV development is the classification and legislation to be applied. The International Maritime Organisation (IMO) has studied the application of rules based on the International Code of Safety for High-Speed Craft (HSC code) which was developed for fast ships such as hydrofoils, hovercraft and catamarans. In 2005, the IMO classified the WISE or GEV crafts under the category of ships.

The International Maritime Organisation recognises three classes of ground effect craft:

Type A: a craft which is certified for operation only in ground effect.
Type B: a craft which is certified to temporarily increase its altitude to a limited height outside the influence of ground effect but not exceeding 150m above the surface.
Type C: a craft which is certified for operation outside of ground effect and exceeding 150m above the surface.

These classes currently only apply to craft carrying twelve passengers or more but as yet nobody has really pushed the WIG, GEV, Flarecraft or any other name for this odd method of transport any further than just concept and military proposals. It is, however, a sibling of the hovercraft.

A futuristic vision of how the ground effect vehicle may take shape with an almost chilling force. (Author's collection)

This illustration offers an idea as to how the WIG could have become a direct rival to commercial shipping, carrying ship-like loads at aircraft speeds with reduced costs. Perhaps we may see the WIG being pushed further again in the future as the economy tightens? (Author's collection)

There are both advantages and disadvantages with GEV. Ground effect craft may have better fuel efficiency than an equivalent aircraft due to their lower lift-induced drag. There are also safety benefits for the occupants in flying in close proximity to the water as an engine failure will not result in severe ditching. However, this particular configuration is difficult to fly even with computer assistance. Flying at very low altitudes, just above the sea, is dangerous if the craft banks too far to one side while turning, or if a large wave occurs then the aircraft will cartwheel violently into the water where it will break up and sink.

A take-off must be into the wind, which in the case of a water launch, means into the waves. This creates drag and reduces lift. Two main solutions to this problem have been implemented. The first was used by the Russian Ekranoplan programme which placed engines in front of the wings to provide more lift (the engines could be tilted so their exhaust blast was directed under the wing leading edge). The Caspian Sea Monster had eight such engines, some of which were not used once the craft was airborne. A second approach is to use an air cushion to raise the vehicle out of the water, making take-off easier. This is used by Hanno Fischer in the Hoverwing (successor to the Airfish ground effect craft), which uses some of the air from the engines to inflate a skirt under the craft in the style of a sidewall hovercraft.

Many people have experimented across the world with the crudest of methods to create a ground effect vehicle, some are as basic as attaching a hang-glider to a homebuilt hovercraft. While the process may work as the pilot strives to become some kind of bird man, deriving from the golden age of aviation where anything was possible and permitted, it can look amazing to watch but it can also be quite dangerous. If the wing was to detach then it would be 'game over' for the wannabe pilot.

TEN

LIGHT HOVERCRAFT

Due to the varied ways in which hovercraft can be operated and the growing interest taken by members of the public, new smaller, lightweight recreational hovercraft started to appear onto the market, craft that could be towed behind a car or stored in the garden. The first commercially available product to enter the consumer market was the Hoverhawk, manufactured by Hover-Air Ltd of Peterbrough, UK, between 1967 and 1971.

This fully enclosed two-seater hovercraft was powered by three Velocette Viceroy air-cooled two-stroke 10bhp engines, two for thrust and one for lift with a maximum speed of 30mph.

The Hoverhawk was the first light hovercraft to go into mass production and 121 were built. Hover-Air Limited was founded by Lord and Lady Brassey of Apethorpe who were pioneers of light hovercraft and Lord Brassey became the first president of the Hover Club of Great Britain, an organisation that has inspired the production of many lightweight hovercraft.

The craft was designed for private, commercial and agricultural use and was supplied around the world to many countries, including Sweden, Mexico, South Africa, Zambia, USA, Canada, Kenya, East Germany, Holland, Singapore and South America. It played a leading role in the 1969 White Nile expedition.

001 was built for evaluation purposes and, although used for demonstration, there were only three Mark 1 craft built. The Mark 2 craft had more or less the same body with more fibreglass than wood. Mark 3 commenced trials registration with CAA as GH-9058 and was a much faster craft due to the Wankel engines it was upgraded with, and a new raised skirt at the bow.

This craft was the first hovercraft to use a moulded glass-reinforced plastic hull, a method of construction which is still in use today for all light hovercraft. Its bag skirt was simple and more stable than those developed for larger craft, quite advanced for the type of craft it was, and still looking modern forty years on.

The Mark 1 and Mark 2 versions used standard motorcycle engines, reducing the production costs and making the craft available at an affordable price. They were the fastest light hovercraft of their era. Controlling the craft was achieved by use of the rudders or by differential throttle making it very manoeuvrable, provided all three engines were working and that was no understatement as it was always early two-stroke engines that would let machinery down rather than the design of the craft itself. Sea water, dust and dirt in electrics can play havoc.

Its gull wing doors made it look more desirable than some of the most expensive super-cars on offer at the time, and in fact it cost nearly as much as an E-Type Jaguar! In an auction in 1975 one was sold for £4,700.

W. G. Spedding. 1969

A DAILY EXPRESS PUBLICATION 5s.

EXPRESS AIR RIDER HANDBOOK

Plans for a new two-seater hovercraft designed specially for school construction.

Details on how to build and how to hover.

Safety regulations, certification of craft, insurance.

What the hoverclub does.

Hoverplaning, fastest growing and newest sport.

EXPRESS AIR RIDER

The *Daily Express* Air Rider became a very popular home-built hovercraft.

Skirt development started with a simple pop-riveted bag skirt made from a type of rubberised canvas dinghy material which proved to have too much friction. Hover-Air Ltd then went on to use Hypolon, a Dunlop material with a nylon weave coating which proved more successful but did not do well on wet sand. The later skirts were made from neoprene-coated nylon with welded seams; this material was excellent on water but built up a large amount of static electricity when operated over wet or long grass. The existing larger craft, operating only on water, did not have the same problems as the lightweight hovercraft, which were aiming to traverse any reasonably flat terrain and the Hoverhawks attempted to overcome a lot of new developmental problems.

One hundred and twenty-one Hoverhawks were built but they were not very stable due to the flat-bottomed hull. They also suffered from being very heavy; most small craft built at this time and later had open cabins to make them lighter. The operation of the craft depended on the reliability of all three engines and the craft were not inherently buoyant. Despite their lack of success the model does represent a significant development in small hovercraft and much work was carried out in experimenting with different skirt materials, some of which work better on land and some which work better on water. Many Hoverhawks have survived and they are important examples of early lightweight hovercraft which not only introduced the use of the craft into a wide number of commercial and agricultural settings but also saw the start of the sporting side of the industry. The Hovercraft Museum has the largest collection of preserved Hoverhawks and all are on display.

Another promising independent hovercraft company was Pindair, from Gosport, UK. Pindair had been marketing their range of semi-inflatable hull-designed craft with the Skima 4 since 1973. The bright orange, four-seater hovercraft was a radical step in the light

An early craft made by Bob Windt of the Untied States.

Mike Pinder's company Pindair offered a range of semi-inflatable hovercraft for consumer purchase. They proved quite durable and were easily maintained and operated by the novice.

hovercraft world: it could be operated by virtually anyone and offered one of the cheapest and easiest ways to get into hovercrafting.

It featured electric start from its single engine and could carry more than its weight at speeds of up to 30 knots. The whole craft arrived at your door in a large box, and all you had to do was add air to inflate it and off you went! But it didn't stop there for this company, as they enhanced their range of products to include slightly larger craft with varied commercial abilities including VIP transport and military applications. The development work carried out by Pindair was closely associated with early Griffon craft.

The simplicity of the hovercraft design means that anybody can build a working example with very limited skills. This is exactly what has happened and it didn't take long before purpose clubs were established in which they would aim to promote amateur hovercraft construction and operation. The most notable club was formed in Great Britain in the 1960s when a group of hovercraft enthusiasts set up the 'Hoverclub of Great Britain'.

From 1960–61 people had begun to construct their own hovercraft from the very little knowledge they had gained, mainly from reading journals published from the time, which included information about all the latest air cushion vehicle developments. Aviation magazines ran regular features devoted to hovercraft technology. These secluded enthusiasts finally had their voice recognised when in November 1964 a statement was made within the 'International News' section of an air cushion publication that a new hoverclub had been established in Great Britain by a group of indivduals from the Isle of Wight and appropriately known as 'The Isle of Wight Hoverclub'. It was in fact the first club of its type anywhere in the world – once again the Isle of Wight and the hovercraft come together in history. It did not take long before the club expanded to such an extent that it outgrew the Isle of Wight! By 1966 the club had changed its name to the Hoverclub of Great Britain and it still remains today as one of the world's most active and friendly hovercraft clubs.

In those early days it was an achievement just to get a 'craft moving and travelling on land and water. That experimentation has not completely disappeared. Many people become active members even if they don't have a hovercraft themselves; taking part in marshalling and lap-scoring are two of the ways that many people become valued members of the

The Hoverlark hovercraft was one of the smallest hovercraft you could buy off the shelf and was designed to fit on a roof rack of a family car! (Author's collection)

Guinea pig – an early home built craft. Most home-built craft of this era were cobbled together from household items and garage leftovers. (HCGB)

club. The Hovercraft Club was financially supported, almost from the beginning, by BP, who gave £3,000 in 1970 purely for administration purposes; therefore the club was able to afford a paid secretary and to produce the early copies of a newsletter. BP also supported the schools 'Build a Hovercraft' competition starting in 1969 which continued for twenty-five years.

The Air Registration Board were responsible for ensuring all craft were safe and met strict guidelines, since hovercraft in this period were considered to be flying and not therefore catered for by any marine agency. Registration was very strict.

The Hovercraft Club has grown over the years and has hundreds of members across the United Kingdom, plus international members who look forward to receiving hovercraft news in the monthly magazine. The club also organises various events which include race meetings, cruises, offshore events and other social activities. Most race meetings form part of the British National Hovercraft Racing Series, and are organised by branches of the Hovercraft Club at various venues around Britain. Hovercraft racing itself is taken very seriously and as such represents a Formula 1 category. The Hovercraft Club has become the controlling body for light hovercraft in Great Britain, and is associated with hovercraft clubs in many other countries round the world.

A European Hovercraft Racing Championship Series is held each year, with rounds in various countries including Britain. Every other year a World Championship Hovercraft Racing Series takes place somewhere in the world. These events bring together international hovercraft racers, designers, mechanics and their families from many nations.

The club also offers the opportunity to take part in International Cruising events including, for example, exploring French rivers. Cruising events are less competitive and allow drivers to explore areas such as coastal mudflats and stretches of river which often can only be traversed in a hovercraft.

Hovercraft racing is not a sport that many have heard of, but most are surprised at the speed and excitement of it. It's the only sport where racing takes place over a mixture of land and water in every lap of each race. A race site could be anywhere from an old gravel pit to the estate of a stately home.

A race will see up to thirty craft on a start grid. When the lights turn green the excitement begins. Races are run by formula, based on engine size:

Formula 1	over 500cc
Formula 2	250–500cc
Formula 503	Standard Rotax 503cc engine and 125kg minimum weight
Formula 3	250cc
Formula 35	Maximum total power output of 35bhp

There is also a Junior Formula for eleven to sixteen year olds, who race in Formula 3-type craft (250cc two-stroke, 500cc four-stroke engines and other limited engine sizes) or Formula 35 craft, giving the opportunity for a parent and child to share the same craft in two different formulas. Most of the craft are currently using two-stroke motorbike, microlight or snow-mobile engines, but there are some competitive four-strokes out there, and Formula 35 uses both four-stroke industrial engines and small car engines (such as Citroen 2CVs or Volkswagens) for reliability.

Formula 1's unlimited capacity makes it the fastest formula, but some of the Formula 2 craft are almost as fast as some of the Formula 1 craft. Both tend to be dominated by craft running separate fans for lift and thrust. Formula 3 tends to use a single engine powering a single fan. The concept behind Formula 35 hovercraft is to have a racing craft that is inherently reliable (hence the four-stroke engine) and is multi-purpose. Formula 35 craft can be used at race meetings or for cruising – they have extra free board built into the hull specially for protection on choppy waters on rivers or inshore. By restricting the power it also tempts drivers to design aerodynamically other parts of the 'craft to increase speed.

A hovercraft race under way at Claydon House organised by the Hoverclub of Great Britain. (Rebecca Taylor)

Hovercraft racing in its early days was challenging to say the least. You can almost feel the fear as the pilot wonders if his craft will make the turn onto land!

A modern Formula 1 hovercraft takes to the air. (Rebecca Taylor)

Any colour you like! (Rebecca Taylor)

Hovercraft racing is a very competitive form of motorsport and there are many different craft. (Rebecca Taylor)

Racing pilots skilfully drift their craft at a race meeting at Hackett's Lake. (Rebecca Taylor)

Formula 35 is generally considered a good formula to start with as the craft are lower powered and it means that you have to learn how to really drive properly – you cannot afford to lose any excess energy.

However, this is not just a sport for wannabe die-hards. Many of today's national and international champions started in racing Junior Formula, which is for youngsters aged eleven to sixteen using Formula 3. As the Junior craft used are the same as in other formulas, those over sixteen can race the same craft in Formula 3, Novices and Open Races, though not in Formula 1 races with a Rotax 447. This is great for parents, schools and youth groups. Formula 1 is restricted to experienced drivers from the lower formulae.

There are many other hovercraft clubs located around the world. The Hoverclub of America is another similar club that promotes the interests of pursuing hovercraft activities for the enthusiast. Then there is the important body that represents the global interest of the hovercraft world. The World Hovercraft Organisation is a not-for-profit organisation whose principle aim is to serve as an information resource for the international hovercraft/air cushion vehicle world. The organisation has no formal membership requirements and is designed to support existing bodies, such as the World Hovercraft Federation, with promotional opportunities, sponsorship efforts and information services.

The primary goal of the World Hovercraft Organisation is to advance hovercraft/air cushion vehicle technology by embracing, expanding and providing communication between all aspects of the hovercraft world. It is a valuable and vital service that supports the hovercraft community of today.

But it's not just amateur and racing craft that have adapted the use of the small hovercraft. Around the world can be found numerous organisations that insist on using hovercraft as part of their vital operations, or in some cases can only operate their activities around having a hovercraft.

The author with his Starbreeze RX-2000 craft which was rescued from a field and brought back to life with the aid of a crash-damaged microlight! (Author's collection)

HoverAid is a unique charity that operates solely from the use of its hovercraft in delivering vital aid and medical supplies to the areas of the world forgotten by international concerns or indeed where the media have failed to report. A lack of access often stops governments, development organisations, NGOs and missions helping communities to develop. It stops communities accessing markets and services that could help them develop themselves and it stops individuals accessing vital health and education services which could quite simply save lives.

HoverAid believe it is hard enough being amongst the majority poor on this planet without being consistently overlooked whenever help is available, because you are simply too difficult to get to.

It is the only organisation of its type in the world and, while UK based, its aims are global: to bring physical medical and spiritual help to save, sustain and enrich human life.

HOVERAID REACHES THE UNREACHABLE

In 1969 Lake Chad was slowly evaporating. The Missionary Aviation Fellowship (MAF) was operating floatplanes on the lake to access over 1,000 islands and 250,000 people. The ever-changing level of the lake placed limitations on the use of motorboats, Land Rovers and MAF's amphibious Cessna aircraft.

Tim Longley, a maintenance engineer and qualified aircraft designer, realised a hovercraft was the ideal solution to negotiate the lake's marshy edges. Whilst on leave at home in the UK Tim discovered there was no suitable hovercraft commercially available and so set about designing the Missionaire (a hovercraft which subsequently held the record for a light hovercraft circumnavigating the Isle of Wight!). It was the precursor to the River Rover, a lightweight, simple, bolt-together design of hovercraft with a revolutionary control system which allowed precise control of an air cushion vehicle along river systems, up rapids, and across swamps, by banking the vehicle.

Despite never reaching Chad, the River Rover design was evaluated by the Royal Navy and adopted for use by the Joint Services expedition to Nepal in 1978/9 which used hovercraft to provide a mobile clinic – or hoverdoctor service – to people living next to the raging torrents and multiple rapids of the Kali Gandaki River as it descended through the Himalayas.

Further expeditions in 1982 to Peru, and 1990 to China, further proved the value of hovercraft as a means of delivering medical services, and opening up remote regions by using river systems as hover highways. In November 1991 HoverAid was created as a charity to support a long-term project in Papua New Guinea where two Mk4 River Rover hovercraft were deployed to extend the reach of Balimo Hospital in the middle of the extremely remote Western Province.

A further expedition as part of a development programme in Nicaragua saw a River Rover operate on the Rio San Juan, and this was followed by another being delivered to operate on the Zambezi of the Barotze Plain in Zambia. It is a sad fact that all of these projects have suffered from lack of infrastructure support and funding, and HoverAid nearly called it a day in 1999. However, in March 2000 reports of cyclones and severe flooding in Mozambique hit the headlines around the world. HoverAid knew there was a perfectly serviceable hovercraft in Zambia and so sent word to supporters in churches

A modern Eagle craft operated by Hovercraft Search and Rescue UK. The charity provides a unique flexible service with the use of hovercraft. (Hovercraft Search and Rescue UK)

The team of volunteers pose for a crew shot on one of their craft. (Hovercraft Search and Rescue UK)

up and down the country. In three weeks they raised nearly £100,000 and moved River Rover 403 to the Save River where, in collaboration with World Vision, it reached 10,000 people stranded on mudbanks without adequate food or shelter.

With further efforts in Malawi in 2001, again during flooding due to cyclones, HoverAid demonstrated that hovercraft could provide crucial access for relief organisations.

Hovercraft Search and Rescue UK (HSR-UK) is a charity that exists purely around having the unique ability to rescue people from areas and situations where other craft would either take too long to respond or, as with many cases, just couldn't get to.

They have two new, fully equipped Eagle 4 hovercraft which will form part of a UK-based flood response team. These craft will allow the team to expand the current areas of operation to include the confines of flooded towns and cities. After the experiences of floods in Windsor in 2003 HSR-UK determined to acquire hovercraft small enough to be manoeuvred around the sorts of obstacles these areas of operation present. These hovercraft, however, only represent the first phase of a planned expansion for the charity and it is anticipated seeing the fleet expand with a variety of different craft to suit different scenarios.

The hovercraft has therefore by no pure intention evolved into many things, but in the field of saving lives it is today a highly valued emergency vehicle. The Eagle 4s are produced by Nottingham-based K&M Products who have been building light craft to their own design for many years, serving the needs of countless amateur builders. Not only does K&M produce ready to fly craft but it also stocks skirts, hulls, ducts, cables, fans and basically anything and everything needed to construct your own hovercraft, either from your own design or theirs. A quick trawl on the internet will result in the many amateur DIY plans available for enthusiasts and schools to make their own hovercraft from plywood.

ELEVEN

HOVERTRAVEL: THE WORLD'S OLDEST COMMUTER HOVERCRAFT SERVICE

On 24 July 1965 the Isle of Wight would once again share another link with its hovercraft partner, this time a direct link that would connect the two, as a new passenger service had opened for business taking pleasure seekers, holidaymakers and commuters across the Solent from Stokes Bay to Ryde.

The first hovercraft was an SRN-6 which went straight into service on the very day it was delivered, making fast use of the new company's asset. And it is this quick method of thinking, a fun approach to business that has helped Hovertravel stay above the competition well into the twenty-first century. While there have been many developments and achievements since 1965, there has always been one constant. The hovercraft remains the fastest way to travel between the UK mainland and the Isle of Wight. It flies directly from shore to shore, transporting its passengers and freight just footsteps from the main road at Ryde in only 7 minutes!

As with most pioneering air travel start ups, Hovertravel had a quite primitive beginning despite having the latest £1 million hovercraft in their livery. If you were to walk down the Southsea seafront in 1965 looking for Hovertravel's state-of-the-art 'swinging sixties' concrete offices and passenger terminal you would have had a wasted time. Nestled between a memorial to one of Nelson's battleships and an amusement park, where you could buy kiss me quick hats and jellied eels, was a caravan and a nice new garden shed that could have housed a mower. (No, not a hover mower – you would have had to wait another twenty years!)

In fact in the early days if you wanted to experience the hovercraft all you had to do was wait in line to buy your ticket from the caravan. This may seem quite quirky but it was the start of a new era in travel and one that has subsequently earned Hovertravel an internationally respected reputation for excellence.

It all started with six individuals who were all so impressed with Sir Christopher's machine that they started to think of a viable way in which they could develop a plan to turn the hovercraft into a business. The partners that formed the company all shared an avid love of sailing, a pioneering attitude and a degree of bravery! These men were Desmond Norman and John Britten, who had founded the Britten-Norman aircraft company at Bembridge; Don Robertson, an outstanding pilot who had flown the empire routes and huge biplanes carrying mail to the Arctic (Don had also been a test pilot for Supermarine where he was one of the first to fly the Spitfire); Edwin Gifford, a civil engineer; Frank Mann, Desmond Norman's business partner in an aerial crop spraying company; and David Webb, a chartered accountant. From this assortment of people we can

The first aluminium-welded hull enters the water for flotation tests. This craft was the first hovercraft to feature a welded hull rather than riveted, with previous craft following traditional aircraft techniques.

clearly see that the basis of the company was formed upon flying backgrounds and also business experts which would both become vital in this pioneering operation.

They all agreed that the new venture would involve a unique passenger service, something different but also welcoming in a change to the usual ferry service. Ryde was carefully studied, being the main passenger port to the island from the mainland. Due to sand banks and low tides, ferry boats have to berth at the end of a half-mile-long pier where their passengers either have to get on a vintage tube train, bus, taxi or walk to the shore; all this takes extra time and in some cases expense.

The pier is still in use today for conventional ferries but back in 1965 Hovertravel bosses seized on the hovercraft ability to travel across both water and land. No matter what the tide was doing, passengers could embark and disembark in exactly the right place, with no need to trundle up and down the pier as an additional requirement of their journey.

And so then the Hovertravel route was born, adapted over the years, but always operating to use all of the unique advantages that the hovercraft has over its rivals.

The company's Senior Captain, and initially its sole pilot, was Peter Ayles, who gave the Governor of the Isle of Wight, Earl Mountbatten of Burma, his first scheduled hovercraft trip when he inaugurated the service on 4 August 1965. Within only a few weeks, the SRN-6 had carried over 30,000 passengers! Hovertravel celebrated with champagne when over 1,000 passengers used the service in one day. Less than six months after the inauguration, 116,000 passengers had discovered the joys of the hovercraft.

The very first timetables did not run strictly by the clock and were quite relaxed as the service was run more on demand until later methods were introduced. Peter Ayles once set an astonishing personal record of seventy-seven crossings in a single day!

Peter Ayles, who had been valiantly boomeranging his way back and forth across the Solent, had his workload eased when two new pilots were recruited, Tony Smith and Peter Atkinson. As highly skilled as they were, they still did their stint on the ticket desk when needed, proving the point that Hovertravel was a business of love rather than a corporate nine-to-five existence.

The fares were 7s 6d for early morning journeys and 10s for crossings made after 10.30a.m. If the hovercraft was not a cause of excitement in itself then what it offered next was sure to heighten the experience when Hovertravel added a crew of new stewards that would make the passengers' crossings as refined as they could be and associating the journey with flying even more so. These hostesses were quickly known as 'Hovergirls' and their uniforms reflected the glamorous fashions of the day.

The Stokes Bay service closed in 1967, but the Ryde–Southsea service became increasingly popular. In 1968, Hovertravel carried its millionth passenger.

The Solent service began to operate a regular timetable and by the end of 1970 had achieved a total of 15,000 running hours, with 17,000 crossings made in that year alone. The company's finances, initially somewhat precarious, improved enormously.

In 1970 a record was set when the SRN-5 achieved a crossing time of just 4 minutes 53 seconds – albeit without passengers, many of whom would have loved to experience this joyride to Ryde.

Another interesting fact: Hovertravel's service was inaugurated in the same year as The Beatles had a hit with Ticket to Ride. While the title of the song does not directly concern the small town on the Isle of Wight, there was perhaps an underlying link that reflects Paul McCartney's visits to the island where his cousin had a bar in Ryde.

It was such a strange thing, a new public service and one that bureaucracy were constantly scratching their heads over as they tried to decide whether the hovercraft was a plane or a ship or even possibly neither. Thankfully though the public regarded Hovertravel and its workings in a humorous capacity while the red tape makers continued to see what obstacles they could throw into the air to make themselves worthy of merit. It is so often the case that such officials can put an end to a good story or crush the intentions and actions of a wonderful organisation without themselves having a clue of what they are really doing. Another personal reason that I admire Hovertravel, along with the millions of other uses of the service, is that they are still as strong and humorous an operation today.

When the hovercraft service came about it was relatively unknown as to who the potential customers would be. At first, visitors to the terminals would note the caravans and maybe think that this was a continuation of the amusement park next door, but once inside the SRN-6 across the Solent their faces would no doubt have expressed a smile far greater than any timid ferris wheel could create. But it wasn't just about fun; the public found the new hovercraft service very useful, not just in cheering everyone up but in providing an unrivalled passenger service between Ryde and Portsmouth. A mix of holidaymakers, day trippers, locals, enthusiasts and commuters made the hovercraft service successful and it is still this rich blend of people that continue to hover across the Solent today.

These customers enabled Hovertravel to invest in updated terminal buildings, although they were still noticeably rather more shed-like than transit lounge. Since then the company's terminals have progressed steadily towards their present state of welcoming with the opening on 3 August 2003 by HRH Prince Philip, the Duke of Edinburgh, of the Ryde terminal.

Hovertravel's sister company, Hoverwork, offered a variety of services, from private charters of its hovercraft fleet to conducting surveys on other continents. The first contract was to supply a hovercraft and crew for an action scene in a film titled *Murderers Row*, which starred Dean Martin and Karl Marden.

Hoverwork are often facilitating hovercraft charters for private trips, including weddings, birthday parties and corporate events. Whilst respecting the environment, destinations range from Osborne Bay to other private and remote beaches.

Another feather in Hovertravel's cap occurred in 1967 when two of its SRN-6 craft were sent to Canada for Expo '67, the international trade fair on an island in the St Lawrence River. Hoverwork ran a ferry service between Montreal and the exhibition site, operating for 12 hours daily and carrying 366,633 passengers during the six months that Expo '67 was open.

In 1968 Hoverwork provided craft and crew for the scientific Amazonas Expedition, organised by National Geographic. The route covered 2,400 miles of treacherous jungle rivers in the heart of South America. The SRN-6 hovercraft left Manaus in Brazil on 11 April and after four stimulating weeks of speeding its way through the wilds of South America, it emerged triumphant in Trinidad on 9 May.

The hovercraft's amphibious abilities have proved invaluable in such projects as the one in which Hoverwork set up a communication network with aboriginal communities in Australia's Northern Territories, where reefs and shallow water prevented access by conventional boat.

But by far the most radical alteration to Hovertravel took place in 1983 when their faithful SRN-6 craft bowed down and retired to make way for a new hovercraft which was to signal the start of a new era in hovercraft. The first hovercraft used on the Ryde-Southsea route were jet-engined SRN-5 and SRN-6 craft that carried between eighteen and thirty-eight passengers. They proved their worth in Hovertravel's early years, but as technology advanced and the service became ever more popular, there were demands for more passenger capacity and less noise, not to mention the rising costs of fuel prices.

Like all technology, time allows changes that improves efficiency. Hovercraft have always advanced and by the early 1980s a radical change was taking place. Gas turbine engines had been the chosen power source for most commercial hovercraft; they offer vast amounts of power and the technical skills needed to build and operate these engines were shared among the aviation industry that built and operated hovercraft at this time. We are all aware how noisy jet engines are when we visit airports. The older they are the more noise they seem to create. While they are quite powerful they have a rapid fuel consumption and are therefore prone to large running costs. However, during the '50s, '60s and early '70s Britain was going through its jet craze. The nation had created the beautiful Comet, the world's first jet airliner, and the descendent of the type, the anti-submarine warfare Nimrod, continued in active service with the RAF until 2011 when defence cuts grounded them. Then came the jet-powered Rover car, which was nothing short of a comical joke. Even a tilting train, the APT-E, was gas turbine powered to 150mph. But things were changing. By the 1980s the hovercraft had proved itself and new cleaner and more fuel-efficient engines were coming into the mainstream – diesel power was here. Out with the old and in with the new. It was the advance of modern diesel engine technology that would spawn the new dawn of hovercraft. In 1983 Hovertravel Ltd introduced a revolutionary new craft into its Solent crossing, the diesel-powered AP1-88. Like the SRN-6 it replaced, the new AP1-88 was designed by BHC and Air Vehicles but built largely by Hoverwork. The most noticeable feature was the twin-ducted fans which reduced noise even further, while eighty passengers could now hover across the Solent. By 1985, Hovertravel's twentieth anniversary, over 8.5 million passengers had travelled on this crossing!

The prototype AP1-88, named *Tenacity*, was launched by Sir Christopher Cockerell in March 1983 and was joined three months later on the Solent route by her sister craft, *Resolution*. This craft was later sold to the US Navy, when a replacement, *Perseverance*, made her maiden trip on Hovertravel's twentieth anniversary. Later on she went to the United States Navy.

The AP1-88 demonstrated the future of hovercraft but for some reason very few orders came in. On the whole the modern AP1-88 was not only quieter, more economical, and could carry more passengers, but more importantly it showed how the hovercraft could adapt to the fast-moving world it worked in. These AP1-88 craft have been built under license in Australia by NQEA in Cairns, operating ferry services in Australia, Spain and Canada. There is also a half-well deck model AP1-88 serving with the Canadian Coastguard which can take a 12-ton payload and has various equipment including a crane to assist with its operations. There were also services running in Norway, Sweden, Russia and Cuba. China also built versions which are still in use today. A mail service in Alaska was also run.

This new craft offered a 7-ton payload and unlike previous commercial hovercraft its hull was welded rather than riveted. The new craft could adapt itself for a wide range of roles including search and rescue, ice breaking, fire fighting and anti-submarine warfare. In a commercial passenger role the craft can take over 100 passengers. In a military personnel-

Three AP1-88s come ashore. (Warwick Jacobs)

The first AP1-88 craft under construction at Cowes, Isle of Wight.

carrying role the AP1-88 will take up to ninety kitted soldiers or two Land Rovers and equipment. Unlike the SRN-6, the AP1-88 had its cockpit area placed on top, similar to the SRN-4 and enabling the pilot to have commanding all-round vision.

	AP1-88
Length	24.4m
Beam	11m
Height on Cushion	9.5m
Weight	29,480kg
Maximum Speed	50 knots

In 1999 Hovertravel's craft were stretched by 3ft in length which enabled the operators to install new water-cooled turbo diesel engines which were 50 per cent more powerful than the air-cooled turbo diesel engines. This new variant now became the AP1-88/100S with four Deutz 12-cylinder engines, two engines for lift and two for thrust.

But it doesn't just stop there for the life of a Hovertravel craft. A hovercraft is quite a valuable vehicle and as such has a global demand for a variety of roles and applications. There is a second-hand market that exists where new owners will put these craft into service, sometimes on the other side of the world or in extreme environments totally different from where that craft may have originally been operating. This highlights the overall versatility of the hovercraft in being able to adapt to a range of work over its life-time. They can navigate extremely hazardous areas, such as stretches of coral reefs bounded by shallow waters, which would be inaccessible to any boat, and unlike other forms of transport, they also have a very low sound signature which does not affect the accuracy of seismic recordings.

Tenacity, Hovertravel's first AP1-88, was on the Solent from 1983 to 1989 before going to Sierra Leone and then on to Canada, where it carried passengers on whale-watching tours.

Tenacity's sister hovercraft, *Resolution* and *Perseverance*, were drafted in by the US Army as military training craft, and later employed in the wilds of Alaska on various commissions, including mail delivery. *Courier* was later purchased by the Cuban tourist authorities, while *Idun Viking* is used for oil industry work within the Caspian Sea.

Other hovercraft, reaching those areas that conventional boats and planes cannot even approach, have gone out to work in the outreaches of Scandinavia, Kazakhstan and Australia.

Liv Viking has worked in Denmark and Canada and one is now used by the Canadian Coastguard for search and rescue work. Renamed *Penac*, it has proved highly effective among the shallows and turbulence of the North American seas and frozen channels.

The latest edition to the Hovertravel fleet is the BHT 130 which again like the AP1-88 is a continuation in modern hovercraft technology. This new hovercraft shares more in common with an airliner than anything before. The pilot has a fully computer-ised glass cockpit and fly by wire control systems. Not only are these craft faster than the AP1-88s but they can carry 130 passengers while coping with a 2m bow wave. Griffon Hoverwork were involved with the BHT and it is very much a modular platform that can be adapted to suit a range of applications from passenger ferry to fire fighting and logistics.

An AP1-88 craft taking pleasure flights across the Solent passes the Hovercraft Museum site on an open day.

Roll out of the first AP1-88 craft for Hovertravel.

A example of the AP1-88's capacity.

In the Solent, the AP1-88 operates within close proximity to where the craft were built at Cowes.

The frozen extremes are no problem for the AP1-88. This example is operated by the Scandinavian Air Service seen here in Denmark during the winter of 1985.

Hovertravel's latest craft, the BHT 130 at Portsmouth. (Hovertravel)

The BHT 130 lifts up at the Ryde terminal. (Hovertravel)

Boarding as a passenger you could quite easily forget you're even on a hovercraft as the powerful diesel engines tick over, sounding and vibrating more like a bus!

The BHT 130's engines each produce over 1,000hp and you'd forget you were moving if you didn't remind yourself by peering out of the window. From inside the noise level is only as loud as riding on a local bus. This is the best example of how civilised a hovercraft is to travel on.

TWELVE

GRIFFON HOVERWORK

In 1976 a new manufacturing business was established from the New Forest in Hampshire, on the same stretch of water where Admiral Nelson's HMS *Victory* was launched in 1765. This new operation would give rise, pardon the pun, to another British hovercraft company, however whilst everyone else had plans of vast gas turbine engine monsters, this new company had a very different aim. Griffon Hoverwork Ltd was the result of a joint effort by a small group of friends that had already partnered together in business with the formation of Hovertravel, the Isle of Wight passenger service, in 1965. With the knowledge they had gained from Hovertravel and their backgrounds in design, construction and operation of hovercraft, civil engineer Dr Edwin Gifford and former Spitfire test pilot Don Robertson were able to set the workshop facilities up at Carlton House near Ringwood, which at the time was little more than a shed in the woods. The close proximity of the River Itchen made the location ideal as it was easy to get out onto the open waters of the Solent.

As early as 1962 Don Robertson had constructed a small craft of his own design. In this same year Edwin Gifford became a consultant to recently formed Hovercraft Development Ltd which was the British Government research body responsible for developing hovercraft technology further. At this time the hovercraft patent was still held by the government and they wanted to explore every route that could lead to revenue for its commercial undertakings.

However, Griffon had a particular objective with their designs and it did not include a business model that they initially thought would be capable of taking on the BHC at East Cowes. As Edwin Gifford and Don Robertson were well aware of the costs involved in operating the powerful SRN-6 craft which used costly and noisy gas turbine engines, it was their idea to produce a craft for a smaller sector utilising an existing power plant but that would still have commercial capabilities. This was the backbone behind the concept of Griffon Hoverwork and it would be one that would eventually earn them the reputation as the world's leading hovercraft manufacturer.

We often hear of so many like-minded stories of success with business, evolving from 'Dream Teams'. Sir Richard Branson built his empire because he had fun with what he did and never took himself too seriously but always surrounded himself with inspiring people that shared his creatively and motivation. Then there is Ron Dennis and Gordon Murray of McLaren who in their spare time decided to design one of the world's most beautifully engineered supercars of all time, the legendary McLaren F1.

In 1971 Edwin Gifford, John Gifford (Edwin's son) and Don Robertson together produced their first hovercraft which used a 4.2-litre Jaguar car engine and featured a convertible roof taken from a sports boat. It was named *Griffon* and ultimately became the forerunner to the formation of Griffon Hoverwork Ltd established in 1976, of which Edwin Gifford would become chairman. The team of enthusiasts had seen success from Hovertravel and now saw the potential for smaller and cheaper craft that would be an alternative to those of the British Hovercraft Corporation.

However it wasn't as straightforward as that. The British Government held the patent for the hovercraft and they did not want to just lend the rights to permit anyone from commercially producing hovercraft. BHC had the major stake in worldwide hovercraft interests while Hovercraft Development Ltd was the body in place that would aim to develop the hovercraft further with possible views of obtaining further patents to the design that the British Government could then release to parties they deemed as fit and responsible. Edwin Gifford was consultant to Hovercraft Development Ltd and as such was approached by Sir Christopher Cockerell for advice on a scheduled commercial hovercraft service. That was how the table was laid for the very beginning of Hovertravel but it also paved a way for Edwin with Hoverwork Development Ltd and with also the government itself for Edwin and his partner's later intentions of producing their own commercial hovercraft.

Up until Edwin and Don Robertson thought about building their own craft, it had been Mike Pinder of Pindair Hovercraft that had been producing small craft, mainly for recreational use, based on his rigid semi-inflatable Skima 4 design. During this early period of the Griffon story, Pinder and the newly forming Griffon team would work together. John Gifford who had shared a great interest in hovercraft with his father Edwin left his job at Ford to join the Pindair Hovercraft Company. Mike Pinder built the Skima 12 model under licence from Griffon.

The first production hovercraft then followed with the introduction of the Griffon 1000 in 1980 which was a ten-seat craft that could carry up to ten passengers whilst making use of a petrol engine. This set the path for the new hovercraft company and firmly placed them on the map. It also proved that there was a strong international market for smaller and more cost-effective craft, fourteen having being sold to different countries already. Griffon were onto a winner and quickly built a solid and reliable reputation which stands to this day.

However, the improvement in their craft didn't stop there; what they did next would truly simplify the hovercraft industry and take it one step further into yet more practical commercial uses. By 1983 diesel engines had come on a long way since the huge and heavy units installed in ships and unreliable trains that were intended to replace steam, although as we know steam would linger long past the conceived timeline that it was meant to. With the electrical engine technology understood new lighter and more powerful turbo diesel engines came onto the market. Griffon quickly saw the potential such engines could have if used with a hovercraft application, for one being more economical than the petrol units whilst delivering more torque. The first craft to receive a diesel power plant was a 1000 model, now designated 1000TD, and was chartered for seismic survey work in China. This craft would also form the basis upon which all subsequent models would follow, although altering in size as the range met different needs. Like other timeless designs, such as the Land Rover, the 1000TD is still in production today after modifications and improvements. Another clear advantage in hovercraft using diesel engines is that they can make do with existing fuel stations at ports seeing as hovercraft spend a percentage of their time at marine locations and other such commercial areas where diesel fuel is commonly available rather than having to draft in gasoline. Some of the most basic changes in hovercraft evolution have come from Griffon, mainly in the use of welding rather than riveting, a method which has been replicated the world over by various manufacturers.

With the ever growing demand for hovercraft Griffon faced a new change in its product range. A larger craft was called for that would meet international demands. In 1987 the decision was made to upscale the range, now that computer technology had made the design process easier and more effective for structural determination while all proven systems gained from previous craft could now be used on the larger variants. The new range of Griffon hovercraft would carry a variety of payloads, spanning from 375kg up to 10 tons and with capacity to carry anything from five to eighty passengers. Griffon also worked hard on new skirt design, including ways to maintain pressure under rough sea conditions with deeper skirts and hydraulic systems that could move larger areas of the skirt to control the air cushion as and when required.

The first in the new range of craft, the 1500TD, was delivered to the Solomon Islands in 1985 followed by another craft to Panama and then the UK for Griffon's own charters. Other 1500TD craft would be pressed into various roles across the world including Pakistan, the Antarctic, Thailand and the USA. And so to the next craft, the 2000TD; with a 2-ton payload it could carry up to nineteen passengers and for an eight-year period three craft entered a commercial passenger hover ferry service along the River Thames in central London where they were granted permission to run at high speeds up and down the river due to being the only craft not to produce wake and very little wash that would cause offence to other river users. The amazing thing about the Griffon London service was that they were operated by traditional old-fashioned Thames watermen who were more used to the slow pace of tugs and barges than fast-moving air cushion vehicles. Not only had they not been in a hovercraft before but they had never flown one and yet they did all of the piloting and servicing of these hovercraft themselves, a testament to simplicity of the Griffon design platform. It has since been worked out that after an average of 10,000 journeys a year in that time the hovercraft achieved an average annual reliability factor of 99.4 per cent.

An upgraded 2000 model craft now called the 2000TDX was ordered by the Swedish Coastguard as it required extra power to negotiate ice ridges in the winter and an increased into-wind and into-wave performance in other periods of the year. To say that the Swedish authorities were pleased with their new Griffon craft was an understatement as the Director General of the Swedish Navy wrote words of appreciation in a letter praising the craft and its manufacture to the British Embassy in Stockholm, shortly after receiving a further two craft. Even today the 2000TDX holds the world altitude record for hovercraft which was undertaken in a journey to the upper areas of the River Yangtze in Tibet led by Michael Cole in 1990; here the craft reached 16,000ft, having flown up from sea level in Hong Kong!

Then in 1993 came one of the most significant orders for Griffon, not so much in terms of volume but more to do with a sense of national pride. Up until this point the MoD had no hovercraft in military service, it had been quite some time since they had disbanded the hovercraft from Naval and Army operations. However, with increasing actions within international theatres and the modernisation of the defence programme the hovercraft would come into play once again with Whitehall. This time it would be the Royal Marines that would quite rightly benefit from the amphibious craft. An order was placed for four 2000TDX craft each with gun mounts on top of the cockpit. They were launched

Hovercraft Development Ltd carried out much work into design, development and operation of future hovercraft. This is the HDL.1 of 1963 which started out as a sidewall craft but later the company built a further example (HDL.2). Much of this experience was subsequently filtered into the rest of the small but growing hovercraft industry.

A Griffon craft in service with the Kuwait Coastguard.

The Royal Marines' Griffon 2400 TDS craft. (Griffon Hoverwork)

Four Griffon 2000 series craft were in service with the Royal Marines until their replacements came into force with the better-equipped 2400 TDS craft. (Griffon Hoverwork)

A Griffon 2000 craft carrying out Arctic survey work. (Griffon Hoverwork)

The 8100 TD series is one of the largest craft Griffon produce; this example is operated by the Swedish Amphibious Battalion. (Griffon Hoverwork)

The 8000 series craft combines decades of hovercraft research and development for a truly cost effective and global answer to many applications. Its skirt system is unique and used an advanced system to maintain its air cushion under the most severe conditions. (Griffon Hoverwork)

AP1-88, known as *Mamilossa*, operating in Canada. (Griffon Hoverwork)

Mamilossa back at the Griffon plant in Southampton undergoing a refit while a new 8000 series craft is under construction close by. (Author's collection)

Ex-Royal Marines 2000 craft waiting for new owners prior to their refits. (Author's collection)

A hoverbarge in operation in a remote condition.

A hoverbarge can provide an ideal answer to transporting large and heavy loads across difficult terrain such as ice.

The HD-1 drifts!

from assault ships while still being air portable and road transportable. The craft provided the Marines with everything they wanted: the ability to move fast from anywhere without restrictions. Perhaps one of their most notable uses occurred during their deployment to Iraq in 2008.

The four craft remained in service with the Royal Marines until 2010 when they were sold back to Griffon who then donated one craft to the Hovercraft Museum and refurbished one for duck hunting in Russia.

However, it didn't end there for the Royal Marines; they went on to receive from Griffon a modern-day hovering fortress. Four newly designed 2400TD models were specially designed with a military purpose in mind from the outset. With a stealth-looking cockpit and armoured control area, this craft is packed with the latest electronic warfare systems, including infrared, radar and gun platforms. As well as being highly manoeuvrable, the craft is faster than the model it replaces and offers a greater payload, obstacle clearance and performance.

An important step in hovercraft history is that they have for some time been classified under the Lloyd's of London shipping registration for their commercial intentions. Griffon are proud to state that they can build their craft to meet the requirements of Lloyd's and indeed any other such classification society. However, the 2400TD is not only utilised for military purposes but ideal for rescue and commercial operations. Similar to the 2000TD, the 2400TD's design allows the user to reduce the width of the craft with foldable side decks, allowable for uncomplicated transportation by road, sea or air.

The Griffon 8000 is one of the largest hovercraft in the range and nine craft are in service with the Indian Coastguard alone. However, the first Indian order presented a bit of an issue as at the time government restrictions on foreign imports meant that if they wanted a Griffon hovercraft then it had to be built in India! After much work in locating a suitable Indian-based firm to construct the crafts a small shipbuilders, Garden Reach in Calcutta, was chosen. The craft arrived in hull form, each with a 40ft container with their engines, fittings, skirts, panels etc. A huge kit of hovercraft parts to assemble in India ticked all the diplomatic boxes and the projects builders were guided by Griffon through the construction process.

In 2000 the Royal National Lifeboat Institution (RNLI) opened up its capabilities to include beach patrols with lifeguards and they also looked closely at how they could tackle the problem of reaching stranded people whom conventional boats couldn't get to in a more cost effective method than deploying air sea rescue helicopters. They considered tracked vehicles and a range of other solutions. In May of that year the RNLI took delivery of an experimental new 450 craft which they planned to evaluate for their needs. It quickly became clear the hovercraft would became a much needed lifeline to the organisation and they placed orders for a further two craft which would operate from the Mersey in Liverpool and from the Wash in Selsey. Today the RNLI operate a total of seven 450 hovercraft in locations across the UK. Three of these craft operate from trucks where they can be quickly taken by road and launched at a location where it would have taken longer to reach by sea and much longer – if possible at all – by lifeboat.

It might seem quite odd that Griffon's work in hovering applies to another area of our social lives that we might not expect: sport. Sports grounds are treated like sacred lands, and as such the grounds at which the celebrity sports personalities perform must be kept in tip top condition at all times, particularly in the prevention of bad weather. Cricket grounds have adopted hovercraft technology with a system used to protect the pitch when rain affects play and to also cover the turf as and when required.

The hover cover is an innovative design by Stuart Canvas that has revolutionised the way that major cricket grounds protect their squares and pitches. Developed in 1998 and installed at Lords in the same year, it not only reduces the amount of playing time lost to rain but also enables the ground staff to cover the pitch and square within 3 minutes, hence reducing the workload of the ground staff. This can be contrasted with the labour-intensive method of covering the ground manually during a rain shower. As quick to remove as it is to install, play can continue on the same pitch conditions as soon as the rain ceases, providing fair conditions and a highly competitive match. The hover cover, with its potential for generating sponsorship and advertising revenue, brings commercial as well as practical benefits to any cricket club, making it an ideal product for covering all international and county grounds.

Griffon, as a company specialising in hovercraft manufacturing, promotes the amphibious capabilities of the hovercraft like no one else, including: hydrographic survey, seismic survey, search and rescue, passenger ferry, civil engineering support, airport crash rescue, mobile medical support, military, paramilitary, charters, private craft, and various other air cushion solutions.

The BHT series, like most of the other Griffon craft, is a modular platform which can be designed to a tailor-made client brief. With a 22.5-ton payload it is by far the most interesting modern hovercraft for engineering platforms where it can be used for transporting vehicles, plant, equipment and fire fighting in remote areas. Griffon have craft operating in some of the most remote and naturally uncomfortable environments on earth, from the freezing conditions of -35° in Alaska and Siberia to the blazing sun of the desert heat in Africa.

GRIFFON HOVERWORK RANGE

380TD
5 passengers
Single engine

500TD
5 passengers
Twin engines

1050TD (Single)
7–11 passengers
Single engine

1050TD (Twin)
7–11 passengers
Twin engine

2000TD
11–22 passengers
Single engine

2400TD
11–25 passengers
Single engine

3000TD
26–42 passengers
Twin engine

4000TD
36–68 passengers
Twin engine

8000TD
54–82 passengers
Twin engine

8100TD
60–98 passengers
Twin engine

BHT130
130 passengers
4 engine

BHT150
150 passengers
4 engine

BHT160
160 passengers
4 engine

BHT180
180 passengers
4 engine

Today Griffon has built over 150 hovercraft which have been sold to over thirty-five nations and they offer one of the largest range of hovercraft in the world. From their huge and historic site at Southampton, the once small operation now employs over 160 people. In 2008 John Gifford and partners faced a new opportunity. They had shaped their company into a modern and successful business with a large team of highly skilled people, but just as importantly, brought the hovercraft into the twenty-first century. Griffon Hoverwork Ltd was sold was advised but to British entrepreneur James Gaggero, Chairman of the Bland Group which has been in business since 1810. I had the pleasure of meeting James whilst on site at Griffon's factory, in fact at the skirt building of all places. James is a successful man who clearly has a huge amount of respect for hovercraft and the men that dared to carry it forward. James agreed that the hovercraft is like so many British creations where we do so little about praising our own efforts or achievements. Griffon Hoverwork started from a leaky shed, went into production looking up to the giant of BHC and ultimately outlived the very company it never set out to rival, thus keeping the spirit of the hovercraft alive today.

THIRTEEN

THE FUTURE

Many people may have thought that hovercraft had disappeared since the demise of the cross-Channel UK service. As we have seen this is not the case as the hovercraft industry is extremely well catered for and is a globally demanded product that many organisations and individuals have come to rely upon. The hovercraft will only continue to grow as an industry as it becomes a more cost-effective way of transporting and competes with the rising costs of air transportation.

There is also another possible direction for the hovercraft in the future. The existing hovercraft as we have to come to know it (i.e. that utilising a crated air cushion), advancing and improving as it does over time, could run parallel alongside a new stablemate. Will this brethren of the hovercraft be a hovercar, flying car, WIG or spaceship? Well it might be the very subject of space itself that this new version will have to familiarise itself with. One thing is for sure and that is that this new craft will have to overcome the natural pull of the earth's gravity.

It must be stressed that physics is the most important aspect in the design and also future role of the new generation of hover vehicles. I see the future being one in which generations from now will witness hover vehicles becoming as commonplace as the SUV. ,

In the 1986 film, *Back to The Future*, an eccentric scientist and his schoolboy nephew take to the skies in a flying car made possible by a gadget called the flux capacitor, which, if it was not bound to the rights of Hollywood and poetic licence, would be the very holy grail of anti-gravity motion, and what could lie beyond that – time travel.

It may all sound like the content of sci-fi news but it is a reality: anti-gravity research is happening as you read this and it is being taken very seriously. As early as the 1950s, the US Air Force experimented with various gravity-defying projects which all seem to tie in around the flying disc accounts of that period. However, you don't have to be a little green man to understand and indeed produce anti-gravity; it is not an invention from another world but one of our own and it is a physical science that we are only just beginning to understand.

Anti-gravity can be produced and demonstrated in a number of ways. The easiest way to explain it is by taking two magnets of the same polarity. If you place one magnet inside a sealed clear tube and then drop the other one on top, the natural reaction is that the two magnets will repel away from each other. The void between the two magnets has a strong forcefield as the top magnet floats – the top magnet is pushing away from the other magnet's magnetic pull and has overcome gravity.

Hovercraft have experienced a similar sort of technology with the hovertrains and many proposals were passed during the early 1980s for mainline maglev (magnetic levitation)

railways throughout the world. However, while various experiments have taken place over many nations, it has been Japan that has seen the most notable progress with their high-speed commuter maglev service that has achieved the highest recorded speed of 361mph, achieved by the CJR's MLX01 superconducting maglev in 2003.

In April 2004, Shanghai began commercial operations of the high-speed Transrapid system. Beginning in March 2005, the Japanese began operation of the HSST Linimo line in time for the 2005 World Expo. In its first three months, the Linimo line carried over 10 million passengers. The Koreans and the Chinese are both building lower-speed maglev lines of their own design, the former at Seoul's Incheon Airport and the latter in Beijing. High reliability and extremely low maintenance are hallmarks of maglev transport lines.

Magnetic levitation is ideal for getting an object to float a few centimetres off the ground but it has its limitations and although elements of its theory are useful I do not feel that it will be the all-powerful force of the future. My personal favourite is that offered by a process known as 'Plasmagnetic Levitation', which seems to be the most promising of all the anti-gravity ideas. It is an experimental technology that involves using ultraviolet radiation to turn air particles into a column of super-conductive plasma. A current is run through this plasma, and a magnet is used to repel against it.

The only major issue involved in turning this from a working wizard's bench test into a Martie McFly hoverboard is the seemly enormous amount of power needed to create the plasma itself. However, this is only the very dawn of this technology; it will be conquered as ideas on generating the required power are found and harnessed within a portable appli-cation. Plasma itself is clean and could definitely have the serious potential to become the energy source of the future.

Hovercraft will be with us for all time.

Above left: Space flight – hovering would be possible on the moon.

Above right: The future looks good for hovercraft!

HoverTech, an exciting company in the United States, has been carrying out research into various anti-gravity technology for some time. HoverTech was established by a group of friends with the desire to turn science fiction dreams of flying cars, hoverboards and cheap spacecraft into reality.

Based near Boulder, Colorado, the HoverTech team has been brainstorming advanced new ways to fly since 1991. Its research team currently consists of five key members plus over a dozen engineers, scientists and entrepreneurs. HoverTech is a think tank, not a research lab. For that reason, the entity does not possess any prototypes based on its theories. HoverTech invites talented individuals to share their visions and ideas so that they can build a platform in which further funding and development can be found under the arm of a dedicated body, all of which makes HoverTech quite unique. They have many interesting theories and potential concepts of hovering vehicles, all of which will make use of a natural source of energy while overcoming the force of gravity. To help finance its operations, HoverTech sells in-depth reports on all of its theories. Since these are phase 1 theories, additional research and development is necessary in order to bring them to proof-of-concept stage.

The elements that make up you and I and everything around us will be included within the power system that makes an anti-gravity vehicle work. No fans, no fossil fuels, no noise, no smell, and an almost endless source of natural energy will be the motor of the future. If anti-gravity ideas, covert military 1950s flying saucer projects and plasma-powered creations seem out of this world, then just wait for what is around the corner.

PATENTS OF SIR CHRISTOPHER COCKERELL

Below is a list of all the patents that Sir Christopher Cockerell filed throughout his career. We know him for his work and achievements on the hovercraft but it would be remiss not to credit and rightly recognise every great technical and social advance its originator brought about. (Source: Hovercraft Museum)

MARCONI PATENTS

Dec. 1935	467996 Improvements in or relating to carrier wave modulator arrangements.
Aug. 1937	500359 Improvements in or relating to navigation aiding radio systems.
Aug. 1937	500481 Improvements in or relating to direction finding radio receiving installations.
Sept. 1937	502972 Improvements in directional aerial systems.
Oct. 1937	504744 Improvements in or relating to navigation aiding radio transmitters.
Jan. 1938	509842 Improvements in or relating to indicator correcting arrangements for use in radio direction finding and other apparatus.
July 1938	517580 Improvements in or relating to radio direction finding receivers.
Jan. 1939	524361 Improvements in or relating to radiogoniometer.
Feb. 1939	526412 Improvements in or relating to radio direction finding receivers.
Mar. 1939	527495 Improvements in or relating to radio direction finders.
Aug. 1939	532417 Improvements in or relating to aircraft radio installations.
Aug. 1939	532418 Improvements in or relating to inductance coil and switch devices.
Aug. 1939	532547 Improvements in or relating to radio direction finders.
Oct. 1939	534945 Improvements in aerial systems for aircraft.

Mar. 1939	536485 Improvements in or relating to radio direction finders.
Feb. 1940	536500 Improvements in or relating to short wave electrical oscillators.
Feb. 1940	539034 Improvements in or relating to electric plug couplings and like connectors or jack plugs.
Mar. 1940	540764 Improvements in radio transmitters.
Feb. 1940	541956 Improvements in electrostatic screens for radio transformers and like apparatus.
May 1940	542223 Improved inductance and capacity trimmer units.
Feb. 1940	542485 Improvements in coupling circuit arrangements in radio receivers.
Aug. 1940	543104 Improvements in remotely controlled selecting devices suitable for spot-wave selection in radio systems.
Jun. 1941	549980 Improvements in rotary electric switches.
Jun. 1941	550619 Improvements in radio bearing indicators and control means.
Aug. 1941	551320 Improvements in graduated scale indicators such as may be used for tuning scales for radio receivers.
Feb. 1940	552497 Improvements in directional radio receiver systems.
Feb. 1940	555968 Improvements in radio directional receivers.
July 1942	558450 A cathode-ray tube suitable for use as an indicator.
Nov. 1942	562032 Holder for piezo-electric crystal.
Dec. 1941	568123 Improvements in radio receivers.
Jan. 1950	683687 Improvements in or relating to navigation aiding radio systems.
Jan. 1949	683688 Improvements in or relating to navigation aiding radio systems.
Jan. 1950	683689 Improvements to navigation aiding radio beacons.
Mar. 1950	683710 Improvements in or relating to radio navigation aids for aircraft.
Mar. 1950	684500 Improvements in or relating to radio communication systems.
Nov. 1950	711273 Improvements in or relating to radar systems.

HOVERCRAFT PATENTS

Some titles have been shortened:

Dec. 1955	854211 Basic air curtain case.
May 1957	893715 Air cushion platform.
May 1957	894644 Landing air cushion for aircraft.
May 1957	895341 Air cushioned aircraft carriers.
Jun. 1958	944501 Side wall vehicle-curtain end seal.
Sept. 1958	935823 Injectors applied to vortices.
Sept. 1958	935824 Pressure induced outboard recovery.
Apr. 1959	919350 Inboard recovery of curtain air.
May 1963	935824 Outboard recovery – to form a second curtain.

May 1963	935826 Outboard recovery – to form a second curtain.
Sept. 1958	935825 Flexible skirt/curtain cushion seal.
Apr. 1959	924496 (With R. Stanton Jones) Recirculation using injectors.
Mar. 1959	944502 Stability and trim control by compartmentation.
Oct. 1959	965748 Variable incidence surfaces for aerofoils.
Aug. 1963	944503 Stability cushions distributed around primary cushion.
Aug. 1963	944504 C.P. Shift.
Jun. 1959	959025 Steering and propulsion.
Mar. 1959	946917 Stabilisation of airflow and prevention of negative lift.
Oct. 1959	959825 Cushion (heave) control for travelling over waves.
Oct. 1959	966135 Recirculation by Coanda effect.
Jan. 1960	968194 Vortices generated by rotating pads.
Mar. 1960	968381 Side-wall vehicle with paddle wheel/air pump.
Apr. 1960	973072 Propulsion by blowing into a cushion.
Apr. 1961	975558 (With D. Hardy) Recirculation–tapering duct arrangement.
Apr. 1960	977060 Recirculation–induced recovery system.
Apr. 1960	975241 Reinforcement of rear curtain by ram air.
May 1960	972068 Hovercraft with inflated side parts.
Apr. 1960	975242 Flexible rod skirt.
Apr. 1960	977061 Positive displacement pump at periphery.
May 1961	983446 Recovery of front curtain air to form a rear cushion.
Aug. 1960	995127 Rail car.
May 1961	997943 Protective air cushion for aerial body.
Jun. 1960	983142 Air bearing.
Oct. 1960	989222 Air pump–fluid brakes.
Jun. 1961	989534 Water separation from recovered air.
Jan. 1962	990745 Inflatable load lifting devices.
Jun. 1961	1002572 Sponge support members.
Jan. 1962	1000771 Controlling flow of fluid by a fluid curtain.
Oct. 1961	1029960 Travelling waves on sidewalls for propulsion.
Dec. 1962	1056070 (With L.A. Hopkins) Flexible wall, inflated parallelograms.
Nov. 1962	1064221 Flexible skirt with coandering fluid curtain.
Dec. 1965	1064222 Inflated skirt, perforated wall, forming skirt deflecting cushion.
Jun. 1963	1073731 Controlled vertical movement of wall to correct roll and pitch.
Jun. 1963	1087734 Flexible wall actuated by fluid flow.
Oct. 1966	1072732 Inflated bag/segment wall.
Nov. 1963	1103191 Guiding means for docking (vertical).
Apr. 1964	1095756 Propulsion by surface effect 'W' (free belts).
Apr. 1964	1110212 Propulsion by surface effect 'B' (discs).
Apr. 1964	1095775 Propulsion, radial members (flails).
Jan. 1965	1092816 Propulsion by modified paddle.
May 1964	1075662 Spray prevention.
Apr. 1965	1135768 (With Messrs Grace & Boutland). 'Boiling water' cushion.
Apr. 1965	1138532 Air-feed flexible duct within cushion space.
July 1967	1236571 Trim control-tapered roller.
Feb. 1967	1216475 Hovertrain-ram air deflectors.

Jul. 1967	1239745 (With D.S. Bliss) Anti-ditch shift of cushion C.P.
Feb. 1967	1219285 A.C.V with wave-top slicer.
May 1967	1228588 Hovertrain-{;one or belt current pick-up.
Oct.1981	1584154 Cushion seal for A.C.V.

MISCELLANEOUS PATENTS

Mar. 1976	1448204 Device for extraction of energy from sea waves.
Jan. 1978	1507916 Energy from sea waves using floats.
Apr. 1980	1571283 Energy recovery equipment from vertical and horizontal water movement using floats

GRIFFON HOVERWORK SALES LIST, 1983–2012

001 1000TD Normandeau, Canada
002 1000TD GSI, China
003 1000TD GSI, China
004 1000TD GSI, China
005 1500TD Solomon Islands Nav. Services.
006 1500TD Terry Little, Panama. Now HoverAid for Madagascar
007 1500TD Mobius GmbH, Germany
008 1000TD WAPDA, Pakistan
009 1500TD ITT Antarctic and USA
010 2000TDX HoverEire, Ireland, Hovercraft Museum
011 2000TD Mobius GmbH, Germany
012 1000TD Royal Thai Navy
013 0000 Not built
014 1000TD Royal Thai Navy, Thailand
015 1000TD Royal Thai Navy, Thailand
016 3000TD Sahaisant, Thailand
017 3000TD Sahaisant, Thailand
018 2000TDX Sea Bus, Tunisia 03/1991
019 2000TDX Sea Bus, Tunisia 03/1991
020 4000TD Triton, India 01/1992
021 4000TD Triton, India 04/1992
022 2000TDX Swedish Coastguard 03/1992
023 1000TD WAPDA 2, Pakistan 03/1993
024 2000TDX Swedish Coastguard 08/1993
025 2000TDX Swedish Coastguard 09/1993
026 2000TDX(M) Royal Marines C21 UK 11/1993 Hovercraft Museum
027 2000TDX(M) Royal Marines C22 UK 11/1993
028 2000TDX(M) Royal Marines C23 UK 01/1994
029 2000TDX(M) Royal Marines C24 UK 01/1994
030 375TD VdF Venice Airport Fire Department, Italy 07/1994
031 2000TDX Finnish Frontier Guard 10/1994
032 2000TDX Finnish Frontier Guard 11/1994

033 2000TDX Infraero, Brazil 10/1994

034 2000TDX Infraero, Brazil 10/1994

035 2000TDX Cole, Nicaragua 03/1995

036 375TD NAU, Brazil 07/1995

037 2000TDX Belgium Army 08/1995

038 2000TDX Landsea Logistics, Nigeria 06/1995

039 2000TDX Finnish Frontier Guard 12/1995

040 2000TDX Airport, Auckland, N.Z. 07/1996

041 450TD Landsea Logistics, Nigeria 08/1996

042 2000TDX Mk II Navy, Venezuela 08/1997

043 450TD Landsea Logistics, Nigeria 02/1998

044 375TD Warsaw Police, Poland 12/1997

045 375TD Galicia State Govt, Spain 05/1998

046 3000TDX Mk II Shell, Nigeria 01/1999

047 3000TDX Mk II Shell, Nigeria 03/1999

048 3000TDX Mk II Shell, Nigeria 06/1999

049 2000TDX Mk II MRCC, Lithuania 06/1999

050 375TD Dundee Airport, Scotland 03/1999

051 2000TDX MkII Border Guard, Estonia 09/1999

052 375TD Min. Ag/Fish, Hong Kong 03/1999

053 2000TDX MkII Border Police, Lithuania 06/1999

054 8000TD(M) Coastguard, India 05/2000

055 8000TD(M) Coastguard, India 08/2000

056 8000TD(M) Coastguard, India 11/2000

057 8000TD(M) Coastguard, India 12/2000

058 8000TD(M) Coastguard, India 07/2001

059 8000TD(M) Coastguard, India 04/2001

060 450TD RNLI, UK 12/2000

061 8000TD(M) Border Guard, Saudi Arabia 12/2001

062 8000TD(M) Border Guard, Saudi Arabia 07/2002

063 8000TD(M) Border Guard, Saudi Arabia 11/2002

064 8000TD(M) Border Guard, Saudi Arabia 05/2003

065 8000TD(M) Border Guard, Saudi Arabia 09/2003

066 375TD Shannon Airport, Ireland 05/2001

067 8000TD(K) KNMPA, Korea 12/2001

068 8000TD Singapore CAAS 01/2003

069 470TD RNLI, UK 12/2002

070 470TD RNLI, UK 13/02/03

071 2000TD Crowley, Alaska, USA 23/06/03

072 470TD RNLI, UK 30/07/03

073 470TD R. Hodson, UK 30/11/03

074 2000TD Navy, Pakistan 03/2004

075 2000TD Navy, Pakistan 09/2004

076 2000TD Navy, Pakistan 04/2005

077 2000TD Navy, Pakistan 07/2005

078 8000TD KNMPA, Korea 11/2004

079 380TD A. Hayward, Senegal 02/2005
080 470TD RNLI, UK 02/2005
081 470TD RNLI, UK 05/2005
082 380TD GHL, Demo craft 06/2005
083 2000TD KNI, Greenland 09/2005
084 2000TD Border Guard, Poland 03/2006
085 2000TD Border Guard, Poland 05/2006
086 8100TD Amphibious Battalion, Sweden 11/2006
087 8100TD Amphibious Battalion, Sweden 02/2007
088 8100TD Amphibious Battalion, Sweden 05/2007
089 1000TD Border Guard, Estonia 01/2007
090 470TD Coastguard, Korea 02/2007
091 470TD Coastguard, Korea 02/2007
092 2000TD Dr J.K. Hall, UK 08/2007
093 380TD Avon Fire and Rescue, UK 01/2008
094 PACSCAT Qinetiq MoD, UK 05/2007
095 470TD Coastguard, Korea 03/2008
096 470TD Coastguard, Korea 03/2008
097 380TD Special Forces, UAE 07/2009
098 1000TD Dianca, Venezuela 06/2008
099 2000TD Peru Navy 09/2010
100 2400TD Royal Marines UK 06/2009
101 470TD Coastguard, Kuwait 03/2008
102 470TD Coastguard, Kuwait 05/2008
103 470TD Coastguard, Kuwait 07/2008
104 470TD Coastguard, Kuwait 08/2008
105 470TD Coastguard, Kuwait 09/2008
106 2400TD Royal Marines UK 06/2008
107 2400TD Royal Marines UK 06/2008
108 2400 Royal Marines UK 06/2008
109 470TD RNLI, UK 08/2009
110 470TD Revamped 470 for KCG 09/2008
111 470TD Morocco – Gendarmerie Royale 05/2009
112 470TD Morocco – Gendarmerie Royale 05/2009
113 8000TD Singapore CAAS 04/2008
114 8100TD Kuwait Coastguard 04/2008
115 380TD Special Forces, UAE 07/2009
116 380TD Special Forces, UAE 07/2009
117 380TD Special Forces, UAE 07/2009
118 380TD Special Forces, UAE 07/2009
119 380TD Special Forces, UAE 07/2009
120 2450TD Swedish Coastguard 09/2009
121 380TD Demo craft 09/2010
124 380TD Merseyside Fire and Rescue 09/2010
125 2000TD Peru Navy 09/2010
126 8000TD Indian Coastguard early 2012

127 8000TD Indian Coastguard early 2012
128 8000TD Indian Coastguard early 2012
129 8000TD Korea Coastguard early 2012

GRIFFON HOVERWORK LARGER CRAFT

AP1-88/80 Hovertravel UK 1983 Bertling, Canada
AP1-88/80 Hovertravel UK 1984 Alaska Hovercraft
AP1-88/80 Hovertravel UK 1983 Alaska Hovercraft
PAP1-88/100 SAS Airline, Sweden 1984 Hoverlines
PAP1-88/100 SAS Airline, Sweden 1985 Bertling
PAP1-88/200 Canadian Coastguard 1987
PAP1-88/100 Hovertravel UK 1989 Sierra Leone airport service
PAP1-88/100 Hovertravel UK 1990
PAP1-88/100 Northern Shipping, Russia 1991 Bertling, Cuba
PAP1-88/100 Hover Mirage, Australia 1986
PAP1-88/300 Comenco, Canada 1987 Hoverlines, Singapore
PAP1-88/100 Taiwan 1994 scrapped
PAP1-88/300 Services International Angola 1995 Hoverlines
PAP1-88/400 Canadian Coastguard 1997
PAP1-88/400 Canadian Coastguard 1998
PAP1-88/400 Canadian Coastguard 2009
BHT 130 Hovertravel UK 2007
BHT 150 Aleutian East Borough, Alaska 2006

INDEX

Air cushion, 12, 13, 14, 17, 19, 20, 28, 29, 36, 42, 43, 47, 48, 54, 56, 57, 59, 60, 66, 72, 93, 110, 112, 116, 129, 139, 143, 149, 150, 166, 169, 174, 177, 181, 182

Air cushion pallet, 13

A.V. Roe, 24

Aérotrain, 54-8, 92

AP1-88, 157-62, 170, 187

Avrocar, 61-4

Bell Aerospace, 60, 72, 73, 110, 112, 114, 115

Bembridge (Isle of Wight), 47, 49, 52, 119, 153

Bertin, Jean, 54, 57, 92, 94

BH-7, 127-32

BHC (British Hovercraft Corporation), 28, 69, 74, 79-104, 157, 164, 165, 176

BHT130, 28, 175

British Rail, 57, 78, 85, 86, 123

Calais, 25, 31, 79, 83, 85, 86, 89, 90, 91, 92, 94, 95, 98, 99, 100

Caspian Sea Monster, 135, 139

CC2, 49-53

CC7, 54

Centrifugal (fan), 19, 30, 31, 81, 117, 127

Channel Tunnel, 94, 99

Cockpit, 12, 31, 42, 65, 88, 96, 122, 128, 159, 166, 173

Cowes, 23, 24, 33, 47, 69, 74, 81, 83, 86, 88, 105, 130, 158, 161, 164

Cushioncraft, 30, 47, 48, 49, 50, 54, 68

Daily Express Air Rider, 41

Dover, 25, 26, 33, 34, 70, 81, 86, 87, 88, 90-2, 95, 96, 98, 100, 102, 104, 105-7

Duct, 25, 29, 30, 31, 63, 69, 110, 125, 126, 152, 157, 182

Ekranoplan , 133-39

Expo 67 (SRN.5/6 section), 74

Falkland Islands, 74, 96

Fan, 19, 25, 29, 30, 31, 33, 49, 63, 69, 81, 96, 99, 110, 111, 117, 121, 125, 126, 127, 133, 145, 152, 157, 179

Formula 1, 145, 147, 149

Goodwin Sands, 88, 109

Griffon Hoverwork, 143, 159, 164-76, 184-7

HDL.1, 167

HM.2, 59

HMS Daedalus, 42, 45, 46, 72, 77, 85, 105, 131

Hover Rover, 66, 67, 68

Hovertrain, 54-8, 177, 182, 183

HoverAid, 150, 152, 184

Hoverbarge, 171, 172

Hoverbed, 14, 60

Hovercar, 9, 60, 61, 63, 67, 177

Hovercraft Club GB, 90, 143, 144, 145, 149

Hovercraft Development Ltd, 164, 165, 167

Hovercraft Museum, 7, 8, 9, 10, 17, 20, 71, 77, 81, 85, 88, 90, 94, 95, 105, 108, 129, 142, 160, 173, 180, 184

Hoverhawk, 140, 142

Hoverlark, 144

Hoverlloyd, 47, 74, 80, 81, 83, 85, 86, 88, 89, 90, 92, 95

Hovermarine, 13, 59

Hoverport, 86, 96, 98, 99, 101, 104

Hoverspeed, 86, 88, 92, 94, 95, 104, 105-9

Hovertravel, 42, 74, 77, 79, 109, 120, 121, 123, 153-63, 164, 165

HSR-UK (Hovercraft Search and Rescue UK), 151, 152

Hydroskimmer, 110, 111, 112

Isle of Wight, 23, 24, 28, 42, 45, 47, 72, 74, 79, 81, 82, 105, 109, 110, 123, 132, 143, 150, 153, 155, 158, 164

Land Rover, 66, 67, 68, 75, 126, 129, 150, 159, 165

LCAC, 60, 112, 113, 114, 115, 116, 117

Light hovercraft, 90, 140-52

Marconi, 18, 117

Mitsubishi, 25

Moller skycar, 63

Momentum curtain, 19, 20, 22, 27, 28

Mountbatten, Lord, 7, 20, 129, 131

N-500, 85, 92, 93, 94

Norfolk, 12, 19, 20, 21

Pindair, Mike, 142, 143, 165

Plenum chamber, 19, 20, 27, 28, 29, 62, 69

Prince of Wales, 85, 92

Princess Anne, 86, 90-1, 94, 102, 104, 105, 107, 108

Princess Margaret, 86-8, 90, 91, 94, 95, 104, 105, 106, 109

Proteus (Rolls-Royce engine), 81, 83, 125, 127

Racing, 28, 29, 145, 146, 148, 149

Ramsgate, 78, 85, 86, 89, 91

RNLI, 173, 185, 186

Russia, 117, 118, 119, 133, 134, 139, 157, 173, 187

Saunders Roe, 23, 24, 25, 31, 32, 33, 36, 39, 42-7, 69, 131

Sea Cat, 99, 100

Seaspeed, 59, 74, 85, 86, 88, 90, 92, 93, 103

SEDAM, 85, 92

Sidewall, 13, 33, 59-61, 139, 167, 182

Sir Christopher, 83, 85, 86, 91, 92

Sir Christopher Cockerell, 7, 8, 17-23, 25, 26, 27, 31, 33, 37, 50, 81, 104, 129, 157, 165, 180

SK-5, 72, 73, 74

Skirt, 7, 27, 28, 29, 30, 34, 36, 49, 53, 54, 59, 62, 63, 66, 72, 83, 86, 88, 92, 93, 94, 99, 100, 110, 121, 125, 126, 127, 131, 133, 139, 140, 142, 152, 166, 169, 173, 176, 182

Solent, 32, 36, 42, 44, 45, 46, 69, 71, 72, 74, 85, 105, 124, 127, 153, 155, 156, 157, 159, 160, 161, 164

Somerleyton, 18, 20

Southsea, 109, 120, 153, 155, 157

SRN-1, 23, 25, 26, 27, 31-36, 42, 45, 47, 104, 127, 129, 131

SRN-2, 42, 43, 44, 45, 69, 131

SRN-3, 42, 45, 69, 131

SRN-4 & Super 4, 28, 33, 39, 79-104, 105, 107, 108, 109, 127, 128, 159

SRN-5/6, 69-78

Sure, 81, 83, 86, 90, 91

Swift, 83, 86, 88, 90, 91

Thornycroft, 17, 119, 124

VA.1, 37, 38

VA.2, 37, 38, 39, 40

VA.3, 37, 38, 41, 42, 45, 79

Vickers Armstrong, 37-42, 66, 67

Vosper, 119-27

VT-1, 120, 121, 122, 123, 124, 125, 127

VT-2, 121, 125, 126, 127

Well deck, 75, 114, 157

Westland, 25, 42, 43, 74, 79

WIG (wing in ground effect), 133-39, 177

Windt, Bob, 137, 142

Zubr, 117-19

If you enjoyed this book, you may also be interested in…

The Hovercraft Story

ASHLEY HOLLEBONE

978 0 7524 6128 1

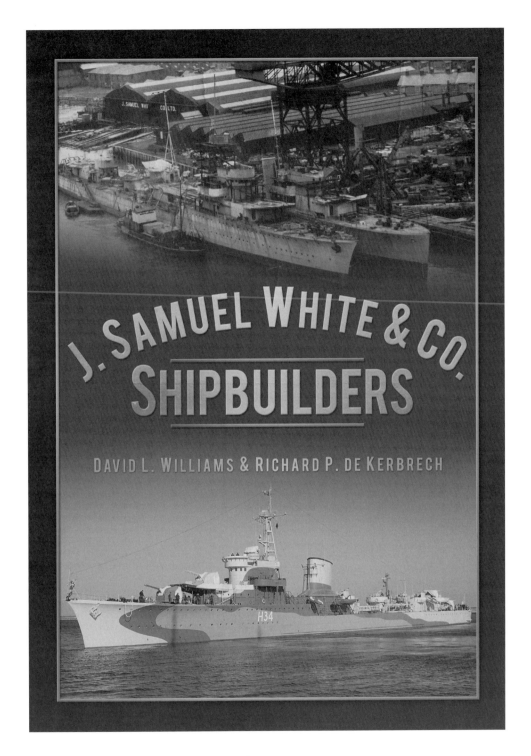

J. Samuel White & Co.: Shipbuilders

DAVID L. WILLIAMS & RICHARD P. DE KERBRECH

978 0 7524 6612 5

Visit our website and discover thousands of other History Press books.

www.thehistorypress.co.uk